METAL IONS IN BIOLOGICAL SYSTEMS

VOLUME 16

Methods Involving Metal Ions and Complexes
in Clinical Chemistry

METAL IONS IN BIOLOGICAL SYSTEMS

Edited by

Helmut Sigel
Institute of Inorganic Chemistry
University of Basel
Basel, Switzerland

with the assistance of Astrid Sigel

VOLUME 16
Methods Involving Metal Ions and Complexes
in Clinical Chemistry

MARCEL DEKKER, INC.   New York and Basel

Library of Congress Catalog Number: 79-640972

COPYRIGHT © 1983 by MARCEL DEKKER, INC.

MARCEL DEKKER, INC.
270 Madison Avenue, New York, New York 10016

ISSN: 0161-5149
ISBN: 0-8247-7038-2

Current printing (last digit):
10 9 8 7 6 5 4 3 2 1

PRINTED IN THE UNITED STATES OF AMERICA

PREFACE TO THE SERIES

Recently, the importance of metal ions to the vital functions of living organisms, hence their health and well-being, has become increasingly apparent. As a result, the long-neglected field of "bioinorganic chemistry" is now developing at a rapid pace. The research centers on the synthesis, stability, formation, structure, and reactivity of biological metal ion-containing compounds of low and high molecular weight. The metabolism and transport of metal ions and their complexes is being studied, and new models for complicated natural structures and processes are being devised and tested. The focal point of our attention is the connection between the chemistry of metal ions and their role for life.

No doubt, we are only at the brink of this process. Thus, it is with the intention of linking coordination chemistry and biochemistry in their widest sense that the series METAL IONS IN BIOLOGICAL SYSTEMS reflects the growing field of "bioinorganic chemistry." We hope, also, that this series will help to break down the barriers between the historically separate spheres of chemistry, biochemistry, biology, medicine, and physics, with the expectation that a good deal of the future outstanding discoveries will be made in the interdisciplinary areas of science.

Should this series prove a stimulus for new activities in this fascinating "field" it would well serve its purpose and would be a satisfactory result for the efforts spent by the authors.

Fall 1973   Helmut Sigel

PREFACE TO VOLUME 16

Of the approximately 90 naturally occurring elements in the
periodic table, nearly all may be found in a human body. However,
fewer than half are known to have any biological function; many are
simply reminders of our geochemical origin. Aside from carbon,
hydrogen, nitrogen, oxygen, sulfur, phosphorus, and the halides, we
need metal ions, though many of them only in trace amounts. Manifes-
tations of the essential trace metal ions are deficiency reactions
if they are supplied in too low quantities; on the other hand, in
many instances toxicity results if the supply is too large. Hence,
it is not surprising that the determination and quantification of
these elements play an important role in clinical, pharmaceutical,
and physiological chemistry, generally speaking, in biochemical
analysis.

The present volume begins with discussions of some nutritional
and immunological aspects of metal ions, followed by considerations
of therapeutic chelating agents used to remove metal ions; 10 of the
18 chapters deal with the determination of metals by various methods.
Another aspect considered is the identification and quantification
of substances like phosphates, cannabinoids, or sulfanilamides via
reactions with metal ions; here a wide and rewarding field of research opens for coordination chemists. The volume concludes with
an evaluation of the chemical aspects in the use of gallium, indium,
and technetium in nuclear medicine.

Helmut Sigel

v

CONTENTS

Chapter 3

THERAPEUTIC CHELATING AGENTS   47

Mark M. Jones

Chapter 4

COMPUTER-DIRECTED CHELATE THERAPY OF RENAL STONE DISEASE   85

Martin Rubin and Arthur E. Martell

Chapter 5

DETERMINATION OF TRACE METALS IN BIOLOGICAL MATERIALS BY
STABLE ISOTOPE DILUTION   103

Claude Veillon and Robert Alvarez

Chapter 6

Chapter 7

Chapter 8

Chapter 12

Chapter 13

Chapter 14

Chapter 18

CONTRIBUTORS

Numbers in parentheses indicate the pages on which the authors' contributions begin.

Robert Alvarez Office of Standard Reference Materials, National Bureau of Standards, Washington, D. C. (103)

R. Bourdon Laboratory of Biochemistry, Hospital Fernand Widal, University of Paris V, Paris, France (245)

Auke Bult Department of Pharmaceutical Analysis and Analytical Chemistry, Subfaculty of Pharmacy, Gorlaeus Laboratories, State University, Leiden, The Netherlands (261)

Clare E. Casey Department of Pediatrics, University of Colorado Health Sciences Center, Denver, Colorado (1)

Jytte Molin Christensen Danish National Institute of Occupational Health, Hellerup, Denmark (185)

M. Galliot Laboratory of Biochemistry, Hospital Fernand Widal, University of Paris V, Paris, France (245)

Raymond L. Hayes Medical and Health Sciences Division, Oak Ridge Associated Universities, Oak Ridge, Tennessee (279)

Kaj Heydorn Isotope Division, Risø National Laboratory, Roskilde, Denmark (123/167/225)

J. Hoffelt Laboratory of Biochemistry, Hospital Fernand Widal, University of Paris V, Paris, France (245)

Karl F. Hübner Medical and Health Sciences Division, Oak Ridge Associated Universities, Oak Ridge, Tennessee (279)

Arne Jensen Chemistry Department AD, Royal Danish School of Pharmacy, Copenhagen, Denmark (139/151/167/185/201/213/225/235)

Mark M. Jones Department of Chemistry and Center in Environmental Toxicology, Vanderbilt University, Nashville, Tennessee (47)

Ole Jøns Chemistry Department AD, Royal Danish School of Pharmacy, Copenhagen, Denmark (201)

Arthur E. Martell   Department of Chemistry, Texas A&M University, College Station, Texas (85)

Poul Persson   Medi-Lab a.s., Copenhagen, Denmark (167/185/201/213/225)

Erik Riber   Medi-Lab a.s., Copenhagen, Denmark (151/167/213/225)

Marion F. Robinson   Department of Nutrition, University of Otago, Dunedin, New Zealand (1)

Martin Rubin   Department of Physiology and Biophysics, Georgetown University School of Medicine and Dentistry, Washington, D. C. (85)

Hans G. Seiler   Institute of Inorganic Chemistry, University of Basel, Basel, Switzerland (317)

Lucy Treagan   Department of Biology, University of San Francisco, San Francisco, California (27)

Adam Uldall   Department of Clinical Chemistry, University of Copenhagen, Herlev Hospital, Herlev, Denmark (139/235)

Claude Veillon   U.S. Department of Agriculture, Human Nutrition Research Center, Beltsville, Maryland (103)

CONTENTS OF OTHER VOLUMES

*Out of print

_____

*Out of print

Other Volumes are in preparation.

Comments and suggestions with regard to contents, topics, and the like for future volumes of the series would be greatly welcome.

METAL IONS IN BIOLOGICAL SYSTEMS

VOLUME 16

**Methods Involving Metal Ions and Complexes
in Clinical Chemistry**

Chapter 1

SOME ASPECTS OF NUTRITIONAL TRACE ELEMENT RESEARCH

Clare E. Casey
Department of Pediatrics
University of Colorado Health Sciences Center
Denver, Colorado

Marion F. Robinson
Department of Nutrition
University of Otago
Dunedin, New Zealand

1. INTRODUCTION

Of the 90 naturally occurring elements in the periodic table, almost
all are present in the human body. However, fewer than half are known
to have any biological role, the rest being fortuitous reminders of
our geochemical origins. The greater part of living matter consists
of only five elements: carbon, hydrogen, nitrogen, oxygen, and sulfur.
These "bulk" elements are present in high concentrations (grams per
kilogram) and daily nutritional requirements are in gram amounts.
The macrominerals--calcium, magnesium, phosphorus, sodium, potassium,
chlorine--are present in the body also in gram per kilogram concentrations and required in the diet daily in gram amounts. These elements may have both a structural and a functional role. The remaining elements occur in the body in much lower concentrations (milli-
grams or micrograms per kilogram). These are the "trace" elements,
originally so called because levels in tissues were too low to be
quantified so that their presence was reported as "trace amounts."
In general, a trace element is one which constitutes less than 0.01%
of the body mass [1,2].

1.1. The Essential Trace Elements

Mertz [1] classified trace elements into two categories: those which
have been established as essential for life or health, and those for
which proof of essentiality does not (yet) exist. Further, an element
may be regarded as essential when a deficient intake consistently
results in an impairment of function and when supplementation with
physiological levels of the element in question, but not of others,
prevents or cures the impairment. To confirm essentiality, it should
be demonstrated by more than one independent investigator in more
than one animal species [1,3].

By these criteria, 15 trace elements are now considered to be
essential in animals (Table 1). Nine of these elements--chromium,
cobalt, copper, iodine, iron, manganese, molybdenum, selenium, and

TABLE 1

Essential Trace Elements

Well-established	Newer	Probable
Chromium	Arsenic	Bromine
Cobalt	Lithium	Cadmium
Copper	Nickel	Lead
Fluorine	Silicon	Tin
Iodine	Vanadium	
Iron		
Manganese		
Molybdenum		
Selenium		
Zinc		

zinc--are well recognized as essential, and most workers would include fluorine. A physiological role has been established for each and deficiencies of all except manganese may occur naturally in animals or humans [4]. During the 1960s, Smith and Schwarz [5,6] developed procedures for maintaining laboratory animals in an ultraclean environment. Using this technique they produced deficiencies, which were later independently confirmed, of nickel, vanadium, silicon, and arsenic. With the possible exception of silicon, no physiological function has yet been discovered to explain the signs of deficiency for these "newer" trace elements [3]. Claims of essentiality have been made for several other trace elements: cadmium, lead, bromine, and tin, but these have not yet been confirmed [4,7]. This list (Table 1) is by no means regarded as finalized.

It may be thought surprising that some elements regarded only as toxic are on the list as being "probably essential." It must be remembered, however, that a number of other elements which are now regarded unquestionably as essential started out as being of concern only for their toxicity. Indeed, for selenium, deficiency has turned out to be a much greater problem than toxicity. Thus, other elements

now regarded as purely toxic or as harmless "contaminants" may be
proved in the future to have an essential role. Such situations
emphasize that no element is inherently beneficial or toxic, but
rather that the biological effect depends on the amount of element
present in the organism. Thus one may have the extremes of too
little or too much causing illness and possibly death, with a range
of intake/tissue concentration in between which is associated with
optimum functioning (health) of the organism.

The essential trace elements are involved in a wide variety of
biochemical functions in the body but most act primarily in enzyme
systems. Metalloenzymes have a very wide range of activity: for
example, there are over 160 zinc-containing enzymes [8] covering all
classes and involved in functions as diverse as gene expression and
maintenance of night sight [9]. Other elements are more restricted
in scope: in higher animals, including humans, there are only four
known molybdenum-containing enzymes, three oxidases and one dehydro-
genase [10], and only one known selenium-containing enzyme, glutathione peroxidase (GSHPx) [11].

Several essential elements have functions in host defense
mechanisms and immune system competence [12]. A number of trace
elements have been found in association with nucleic acids and it
is thought they may act to enhance the stability of DNA and RNA, as
well as of various proteins. Copper, iron, and vanadium can act as
respiratory carriers, although only iron has this function in humans.
Other elements may be of structural importance in the body, e.g.,
silicon in cartilage, fluorine and possibly zinc in bone. Trace
elements also function as an integral part of a vitamin (e.g.,
cobalt in vitamin B_{12}) or hormone (e.g., iodine in thyroid hormones).
Although not part of the hormone molecule itself, zinc and chromium
appear to be involved in the production and functioning of insulin
and zinc may also be important in the activity of other hormones.

1.2. Nutritional Aspects

Table 2 outlines some metabolic and nutritional aspects of the
essential trace elements. In adult humans, the total body content

TABLE 2

Nutritional Aspects of the Essential Trace Elements in Humans

Element	Amount in adult body	Recommended daily dietary allowance	Functions	Deficiency
Chromium	2 mg	50-200 µg[a]	Maintenance of normal glucose tolerance	Disturbances in glucose, lipid metabolism, peripheral neuropathy
Cobalt	1.5 mg	3 µg vitamin B_{12}	Erythropoiesis	Only as vitamin B_{12} deficiency
Copper	80 mg	2-3 mg[a]	Ceruloplasmin, enzymes in synthesis of cartilage, bone, myelin	Anemia, skeletal defects, neutropenia, neurological defects
Iodine	11 mg	150 µg	Thyroid hormones	Goiter
Iron	3-5 g	10-18 mg	Heme respiratory carrier, enzymes, immune system	Anemia, tiredness
Manganese	1 g	2.5-5 mg[a]	Enzymes in protein and energy metabolism, mucopolysaccharide synthesis	Not known in free-living subjects
Molybdenum	10 mg	150-500 µg[a]	Enzymes in metabolism of xanthine, sulfites, sulfur amino acids	Defective metabolism of xanthine, sulfur
Selenium	6-12 mg	50-200 µg[a]	Glutathione peroxidase	Muscle weakness, congestive cardiomyopathy
Zinc	2-3 g	15 mg	Enzymes in most major metabolic pathways, nucleic acid and protein synthesis, immune system	Growth failure, skin lesions, loss of appetite, delayed sexual maturation
Arsenic	1-2 mg	--	Iron metabolism	Not known
Nickel	10 mg	--	Nucleic acid, lipid metabolism, iron absorption	Not known
Silicon	2-3 g	--	Structural component of connective tissues	Not known
Vanadium	15 mg	--	Lipid metabolism, regulation of cholesterol synthesis, ATPases	Not known
Fluorine	3 g	1.5-4 mg[a]	Structural component of bones and teeth	Dental caries, osteoporosis?

[a]Estimated safe and adequate daily intake.

Source: Compiled from Refs. 1, 13, 14, 53, 61.

ranges from a high level of 3 g for iron and zinc down to 1.5 mg for
cobalt [13]. Recommended daily dietary allowances (RDA) have been
formulated for iodine, iron, and zinc based on requirements for
growth and maintenance, fecal and urinary excretion, and absorption
[14]. Cobalt is required only in the form of vitamin B_{12}, for which
a daily requirement and RDA have been estimated. Information about
the metabolism or requirements of other essential elements is incomplete. However, the U.S. National Academy of Sciences [14] has published ranges for the daily intakes of copper, manganese, fluoride,
chromium, and selenium which may be regarded as safe and adequate.
These figures are based on the usual intake from a Western-type diet
by an adult who is healthy and therefore assumed to be obtaining
sufficient quantities of the trace element.

Deficiencies and toxicities of trace elements have long been
known in agriculture where they can be of vast economic importance
[2]. Indeed, most of the early interest and research in trace element nutrition and metabolism arose because of problems with farm
animals. Copper, fluorine, iodine, iron, manganese, molybdenum, and
zinc have all caused problems in agricultural practice [2,15]; the
essentiality of cobalt and of selenium was first recognized from
deficiencies in cattle and sheep. In human nutrition, however, the
situation has been somewhat different. The roles of iron deficiency
in anemia and iodine deficiency in goiter were recognized early last
century, although both diseases have been treated as such empirically
for millenia. Apart from these two cases, it was long felt that man
was sufficiently far up the food chain to be protected from deficiencies of other elements and that requirements were so small that
a deficiency was highly unlikely to occur. The discovery, in the
early 1960s, of widespread human zinc deficiency in the Middle East
put an end to this overly optimistic state of affairs [16]. Since
those early reports of a zinc-responsive nutritional dwarfism, cases
of both acute and chronic zinc deficiency have been recognized in
many parts of the world [17]. Human deficiencies of selenium and
copper have also now been reported [18,19].

TABLE 3

Some Etiological Factors in Trace Element Deficiencies

Factors	Etiology	Reported cases
Inadequate dietary intake	Protein-energy malnutrition	Zinc, chromium, iron, selenium, copper
	Poverty, ignorance	Zinc, iodine, iron
	Low environmental levels	Iodine, fluorine, selenium
Decreased bioavailability	High fiber/phytate	Zinc, iron
	Infant formulas	Zinc, iron
Decreased absorption	Malabsorption syndromes, steatorrhea	Zinc, iron, (cobalt)
	Chemical interferences	Copper
Excessive losses	In urine, sweat	Zinc
	Surgery, burns	Zinc
Increased requirement	Rapid growth	Zinc, iodine
	Pregnancy, lactation	Zinc, iron, iodine
Reduced stores	Prematurity	Iron, copper
Iatrogenic cause	Chelating drugs	Zinc
	TPN, synthetic diets	Zinc, copper, chromium, selenium, molybdenum
Genetic defect	Acrodermatitis enteropathica	Zinc
	Menkes' steely hair disease	Copper
	Xanthine and sulfite oxidase deficiencies	Molybdenum

Apart from inadequate dietary intake, trace element deficiencies may arise from a variety of causes, some of which are outlined in Table 3. To date, deficiencies have been reported in humans for chromium, copper, fluorine, iodine, iron, molybdenum, selenium, zinc, and one possible case of manganese deficiency [20]. It is well known that iron absorption can be greatly altered by the source of iron and

the composition of the diet [21]. This is true also for zinc;
availability of both iron and zinc, and possibly other trace
elements, is decreased in diets composed mainly of cereals and
legumes compared with diets containing some animal protein. This
decreased bioavailability is an important etiological factor in the
occurrence of iron and zinc deficiencies, especially in Third World
countries [22,23]. Decreased absorption due to malabsorption syndromes is frequently associated with trace element deficiencies
(iron, copper, and zinc) on an individual basis. Acute deficiencies
of chromium, copper, molybdenum, selenium, and zinc have all been
reported in patients receiving long-term total parenteral nutrition
unsupplemented with trace elements [24]. Genetic defects in trace
element metabolism resulting in deficiency syndromes [25] have been
reported for copper (Menkes' steely hair syndrome), iron (congenital
atransferrinemia), molybdenum (xanthine and sulfite oxidase deficiencies), and zinc (acrodermatitis enteropathica). Although rare,
such conditions can often provide useful information on the metabolism and functioning of the trace element involved.

2. CURRENT PROBLEMS

There has been an enormous increase in interest in all aspects of
trace element biology in the last 10-15 years, as witnessed by the
number of conferences, workshops, and publications devoted to both
general and specific aspects of trace element biochemistry, nutrition,
metabolism, and analysis [26-33]. Various factors have contributed
to this growth from time to time but probably the most important
include:

1. The development of improved analytical techniques. The
introduction of atomic absorption spectrophotometry in the late 1950s
was undoubtedly one of the most important advances in trace element
analysis. Recent developments, such as the graphite furnace and
microprocessor controlled data acquisition, which allow the analysis
of newer trace elements at very low concentrations, ensure that this

technique remains very widely used. Other techniques, such as
various types of emission, fluorescence and x-ray spectroscopy, and
neutron activation analysis, are also being continually improved to
allow measurement of low levels of trace elements in a variety of
complex biological matrices [34].

    2. Discovery of a function. Many trace elements are accepted
as nutritionally essential before much is known about their biochemical or physiological functions in the body. The essentiality of
selenium was recognized in 1957 and it was known the element was
involved with vitamin E in the mechanisms protecting intracellular
structures against oxidative damage. However, the actual role of
selenium as an integral part of the enzyme glutathione peroxidase
was only discovered in 1972 [35] and since that time there has been
a vast increase in research on all aspects of selenium biology [28].

    3. Recognition of a deficiency syndrome. More interest will
obviously be taken in an element with proven nutritional problems
such as zinc or iodine, than in one such as manganese for which no
naturally occurring human deficiency has been recorded.

2.1. Analytical Problems

In spite of all the technological advances in trace element analytical capabilities in the past 10 years, problems in the detection
and measurement of trace elements have by no means been overcome.
Not unexpectedly, the apparent levels of some elements which are
normally present in blood in very low concentrations have declined
as analytical precision and limits of detection have improved [36].
Given the sophistication and capabilities of modern instrumentation,
of far more practical importance is the environment in which a sample
is obtained and prepared for analysis, especially in consideration of
the ultratrace elements, e.g., chromium, nickel, manganese, and vanadium. It is of little use to be able to measure the concentration
of an element precisely and accurately if all one is measuring is
the level of contamination.

The analysis of trace elements in a biological sample involves a number of steps:

1. Collection.
2. Transfer of sample to laboratory.
3. Storage.
4. Preparation for analysis. Most techniques require some preliminary preparation, such as drying or destruction of organic matter.
5. Measurement.

The analyst is generally aware of the environmental problems at the laboratory end, affecting the sample at steps 3, 4, and 5. Contamination control in the laboratory includes the use of a laminar-flow, class 100 work surface, 18-megohm deionized water, ultrapure reagents, acid-cleaned glassware, and substitution of plastic for other materials where possible. Specific precautions may also be required for individual elements, e.g., the avoidance of glass in chromium analyses, of stainless steel in analyses for chromium, nickel, or manganese. Unfortunately, contamination is more likely to occur before the sample reaches the analyst, at steps 1 and 2, over which the analyst has little control. Nonetheless, he must be aware of the details of the process by which a sample is obtained so that he can judge the ultimate value of his analytical result.

Because "normal" values for trace element levels in a given matrix vary widely from laboratory to laboratory, it is often difficult to make comparisons of results reported by different groups. The increased use of standard reference materials should overcome this problem to some extent, at least with respect to the actual analysis [37]. Unfortunately, there is not yet a reference material for trace elements in the most widely analyzed matrix, blood serum.

2.2. Metabolic Functioning

One of the major areas of concern in trace element research is that of explaining the clinical and pathological changes seen in deficiency

in terms of the biochemical functioning of an element. For some
essential elements, e.g., iodine, copper, silicon, the known biochemical and physiological roles can be reconciled. For most, however, there are large gaps in our understanding and a number of
factors which must be taken into account in interpretation of their
metabolic role [4,7,38].

Deficiencies of most elements produce both specific and nonspecific clinical effects. In experimental animals, all lead ultimately to growth failure, and most also cause various reproductive
difficulties. Thus, even for elements such as copper and manganese,
where specific effects are explainable in terms of the known biochemistry of the element, the nonspecific effects frequently cannot
be linked to a biochemical lesion. This is particularly the case
with the newer trace elements (nickel, vanadium, lithium, arsenic)
for which little is known of the biochemical functioning.

For some elements, a major experimental drawback arises in
that there are species differences in the response to deficiency.
For example, as an integral component of GSHPx, selenium is one of
the mechanisms protecting cellular components from oxidative damage
[11]. When the selenium intake is inadequate, levels of selenium
and GSHPx decline in the blood and tissues but metabolic and pathological lesions are often difficult to interpret. In ruminants,
deficiency causes degenerative changes in muscle cells leading to
muscular dystrophies. In nonruminant species, such as rats and pigs,
it is the liver and heart which are primarily involved. Not all GSHPx
activity appears to be selenium-dependent. The relative distribution
of selenium-dependent and selenium-independent GSHPx varies among
tissues in different species, which may account in part for the
species differences in pathology of selenium deficiency. Several
other nutrients, e.g., vitamin E and sulfur amino acids, are also
involved in antioxidant protection [39] and thus the effects of
selenium deficiency will be modified by the status of these nutrients.
A current question in human nutrition is: why do some groups of children in Germany and New Zealand who have low blood levels of selenium,

comparable to levels in Chinese children with Keshan disease, not
show similar clinical problems [18,40-42]?

The relative timing of events can be important in interpreting
the relationship between biochemical and clinical events in trace
element deficiency [38]. For copper, as for selenium, the first
response to a deficient intake is a gradual fall in level of the
element in the tissues, followed by a fall in the activities of the
active form of the element (usually an enzyme or enzymes). It is
only after this stage that pathological and then clinical lesions
appear. Zinc deficiency, in particular, does not follow the logical
sequence of events that some other elements show. An inadequate intake of zinc results in a decline in activity of some zinc-containing
enzymes, e.g., carboxypeptidase A and alkaline phosphatase, which is
followed very rapidly by the onset of clinical symptoms. The activity
of other enzymes, e.g., dehydrogenases and carbonic anhydrase, is
initially maintained and does not start to decline until some time
after the onset of growth failure.

Indeed, of all the essential trace elements for which deficiency
syndromes have been observed in humans, zinc is undoubtedly the most
problematic. The requirement for zinc is highest during periods of
rapid growth and thus the groups most susceptible to zinc deficiency
are young children, adolescents, and pregnant women. The earliest
signs of chronic zinc deficiency are growth failure and depression
of appetite; subsequently reproductive difficulties may occur and
prolonged, severe deficiency causes lesions in skin and other epi-
dermal tissues. In acute zinc deficiency, as occurs in patients on
total parenteral nutrition, skin and hair lesions appear very rapidly.
Much work has been done on the biochemistry of zinc and its role in
various enzymes. However, as yet, the main features of deficiency--
the very rapid response in growing infants and children and in tissues with a high synthetic activity--cannot be explained adequately
in biochemical terms [38].

2.3. Measures of Nutritional Status

An individual's requirement for nutrients, including trace elements, depends on a wide variety of host factors such as age, sex, health, and lifestyle. Thus measurement of the intake of a nutrient alone gives no indication of the adequacy of the diet in meeting those requirements. A measure is also needed of the nutritional status of that nutrient in the individual, i.e., the amount of the nutrient in the body which is available to carry out its appropriate functions. The parameters used to measure nutritional status must be practicable in terms of both sample material and analytical technique, and must reflect the functional levels of the element in the body. Unfortunately, both conditions are rarely fulfilled. The sample material chosen will depend to some extent on the requirements of the investigator; rapid screening of a large population must utilize a much simpler technique than a detailed research project on a small number of individuals.

The most commonly used measure of nutritional status is the level of trace element in the blood, either whole blood or some fraction. This is a relatively accessible sample and for some elements is easily analyzed, although there are still technical difficulties with others. For copper and zinc, blood concentrations are lowered in deficiency; however, they do not necessarily closely reflect levels in other tissues or functional availability. There are many other factors besides nutrition which affect the levels of these two elements, including infection, hormones, pregnancy, and sampling time. Other body fluids (urine, saliva, sweat) and accessible tissues (hair, placenta) have varying degrees of usefulness. In general, a combination of substrates is more useful than a single measurement.

Analysis of tissues at autopsy is obviously useless for measuring individual status in the living but may give some indication of the general level of trace element nutrition in a population

group [33]. Tissue samples can be obtained from the living during
surgery or by needle biopsy techniques. Such samples, usually
muscle, liver, bone, or skin, are of limited value, as they are
generally very small and thus readily contaminated and may not be
representative of the whole organ. Also, such techniques cannot be
routinely applied to healthy people.

There are two main assumptions in this approach to measurement
of nutritional status: first, that levels in the tissue chosen
(usually blood) reflect the level of the element at the tissue and
cell sites that are sensitive to depletion, and second, that a lowered content of element in a gross tissue sample reflects the presence of a metabolic lesion [43]. The validity of the first assump-
tion varies with the different elements. Blood levels of selenium
do appear to reflect selenium intake and possibly total body content
[44]; tissue stores of iron and copper may be estimated not from the
total concentration of element in the blood but from the level of a
particular molecular species (ferritin and ceruloplasmin, respectively); zinc concentration in blood is not related to levels in
tissues but rather to the rate of mobilization of the element from
the tissues. There is no evidence for any of the essential trace
elements to link low total levels in tissues with a relevant biochemical disturbance [4].

Thus a more desirable measure of nutritional status would be
one which tests some metabolic function of the element rather than
just analyzing the total amount present. Similar criteria apply in
choosing the most appropriate parameters; the function must reflect
the amount of element available in the body, it must respond to
changes in supply, and alteration in the function must be relevant
to the pathology of deficiency or excess.

For an element such as iodine which has one well-defined major
function, a functional test (thyroxine binding) is readily devised
and is sensitive to the status of the element. The selenium-
containing enzyme, GSHPx, reflects the amount of selenium in the
diet and blood where the intake is low, but we do not yet know the
relation of GSHPx to metabolic and pathological lesions of selenium

deficiency. At the other end of the scale, while some zinc-
containing enzymes show reduced activity in response to an inadequate dietary intake, others do not, making the choice of function
measured very important. On a more practical note, the main sites
of action of trace elements, and thus the tissues most immediately
sensitive to deficiency, are rarely those which are readily accessible to the investigator. A promising newer approach, especially for
zinc, is to measure some end product of a process depending on the
element rather than the enzyme involved. Frequently, an end product
such as prealbumin or retinol-binding protein is manufactured at a
different site from that where it is utilized. Thus levels of this
end product in the circulation will reflect availability of an element at its functional site better than circulating levels of the
element itself.

As with all other nutrients, the biological response to a
trace element depends on the supply. For most of the essential
trace elements, there is a range of tissue levels compatible with
optimum growth and function (health). As exposure, and thus tissue
levels, increases or decreases, functioning gradually declines until
clinical toxicity or deficiency occurs. Diagnosis of the overt syndromes of toxicity or deficiency is relatively straightforward and
readily supported by available measures of nutritional status. However, it is the intermediate stages that are becoming of increasing
interest and concern to nutritionists. Nonoptimal intakes (too much
or too little) of a nutrient may not cause overt problems but in the
long term may give rise to biochemical lesions which in turn may
compromise normal growth and development. Thus, research in trace
element nutrition must be directed toward elucidating techniques
which will allow identification of subclinical, pathological changes
in nutritional status.

3. DIRECTIONS

In spite of the many gaps in our knowledge of very basic aspects of
the metabolism of trace elements, their practical significance in

human nutrition is steadily becoming widely accepted. The identification of specific problems with many elements has stimulated interest in possible wider, nonspecific roles of trace elements in human health. One example of this is research into the possible involvement of trace elements in the etiology of the "Western" degenerative diseases, including cardiovascular diseases and cancers. This section outlines some areas of current research in trace element nutrition.

3.1. Interrelationships

Nutritionists have been aware for many years that the level of one nutrient in the diet may influence the requirement or affect the utilization of another nutrient (e.g., niacin--tryptophan; calcium--phosphorus--vitamin D). That nutrient interactions can also affect trace element nutrition is well appreciated among those working in nutrition of domestic livestock [2,15].

A wide range of trace element interrelationships has been observed under experimental conditions [32,33]. Such relationships can exist between two elements, but more likely will be multifactorial and very complex [32,45,46]. The most important interactions from a practical viewpoint are those which occur at the level of intestinal absorption.

Interrelationships may also manifest themselves as a competition for binding sites on transport or storage molecules or substitution at an active site of an enzyme, or as a requirement of one element for the proper metabolism of another. Antagonistic relationships acting at the level of absorption or cellular metabolism can readily be explained from the chemical properties of the elements themselves [47]. Elements with a similar electronic configuration and ionic radius may compete for binding sites, e.g., zinc and cadmium in metallothionein, and magnesium/manganese substitutions at enzyme active sites. Synergistic interrelationships, where one element enhances the role of another, are less readily explainable.

The interaction between copper and iron arises from the role of ceruloplasmin in the mobilization of iron from body stores. Until much more is known about the functioning of other elements, however, we can neither predict such interactions nor explain the biochemistry of those which have already been observed. A general guide may be that one element takes part in some metabolic process in which another element becomes involved at a later stage. There is a third type of interrelationship which arises directly from interaction between the elements themselves. This is best observed in the protective effect exerted by selenium against toxicity due to some heavy metals. Selenium has a high affinity for such elements as mercury, cadmium, and silver, and will form complexes with them in vivo, causing a decrease in the biological activity of both the selenium and the heavy metal.

Although the bulk of reported trace element interactions have been produced under experimental conditions, there are a number which are of considerable practical importance in human nutrition. The essential role of copper in iron metabolism was the first such recognized. One of the symptoms of copper deficiency in infants is an anemia, unresponsive to iron supplementation, but which resolves when adequate copper is supplied [19]. Conversely, an excessive intake of zinc may cause a secondary deficiency of copper. Several such cases have been reported in patients receiving large zinc supplements, including an infant with acrodermatitis enteropathica and adults with sickle cell anemia. Such cases emphasize the concept that for trace elements, like other nutrients, not only is the amount supplied in the diet important but also its relation to the amounts of other elements.

Besides element-element interrelationships, trace elements also interact with other nutrients [32]. Probably the most well known of these interactions is that of selenium with vitamin E. This appears to be a purely functional relationship, both elements playing a similar biochemical role, but probably not interacting directly [39]. Zinc is required for the proper metabolism and function of vitamin A. The strong redox potential of vitamin C may

alter the valence of elements such as copper or iron and thus reduce
or enhance absorption. Undoubtedly many other vitamin-trace element
interactions will be reported, as well as relationships involving
macrominerals, e.g., zinc with calcium, phosphorus, and vitamin D.

3.2. Bioavailability

In order to translate the metabolic requirement for a trace element
into a recommended dietary intake, one must know the bioavailability
of that element from a food or composite diet. The concept of bioavailability includes both the availability of the element for intestinal absorption and its subsequent availability for utilization in
its normal metabolic pathways in the body. For many elements, only
the first part (amount absorbable) has any practical importance.

It is well known that iron is absorbed to a different extent
from different types of foods [21,22]. The bioavailability of iron
depends mainly on the form in which it is present, heme iron being
better absorbed than the form in plant foods or the inorganic salt.
A number of factors present in the diet, such as sugar and vitamin C,
may enhance its bioavailability, whereas others, such as phytate and
tannic acid, may decrease iron absorption.

The bioavailability of zinc and of copper is also markedly
affected by the composition of the diet and the source of the element
[23,48]. Both intrinsic and extrinsic zinc are better absorbed from
animal protein foods than from plant-based foodstuffs. Cereals and
other plant foods contain a number of substances, most importantly
phytate, which can bind zinc strongly, making it unavailable for
absorption. Little is known about the form of zinc in other foods.
The availability of zinc from human milk is much higher than that
from cow's milk and infant formulas but it is not certain whether
this is due to the presence in human milk of a specific enhancing
zinc-binding ligand or whether cow's milk contains substances which
make zinc less available.

The forms in food of selenium [49-52], cobalt (only utilized as vitamin B_{12}), and possibly chromium affect the bioavailability of these elements for absorption, but little is known about the other essential trace elements. Apart from dietary factors there are many host and environmental factors which can influence bioavailability [48,52]. Moreover, decreased bioavailability is one of the main causes of deficiencies of iron and zinc which are widespread, especially in populations whose diet is largely cereal-based. This emphasizes the need for more research into the wider aspects of bioavailability of these and other trace elements, such as their functions in immune systems.

3.3. Speciation and Localization

In recent years there has been a growing interest in the "subdivisions" of an element. Information on the valence state and coordination chemistry of an element in different situations, its binding to intra- or extracellular components, the distribution of an element within a tissue or cell may help in elucidating the biochemical function and metabolism of a trace element and in explaining lesions which may occur in deficiency or excess. Toxicologists have long recognized that the chemical form or valence state of an element can affect its metabolic behavior in the body. The classic case in point is that of arsenic; As(V) is relatively harmless but in the trivalent state, as in As_2O_3, it is the highly toxic substance beloved by generations of mystery writers.

Radioactive tracer techniques provide information on uptake and excretion of elements by the whole body and the pattern of distribution to the various organs and tissues. Biochemical techniques coupled with microanalytical methods of elemental analysis now permit investigation of the distribution at subcellular and molecular levels. As well as chemical form, a variety of factors can influence the distribution of an element in the body, including amount ingested, previous level of intake, presence of body stores, route of excretion,

site of principal metabolic functions, and many host factors, such
as disease processes and hormonal status.

Postmortem analysis of tissues provides some description of
trace element distribution in the body but gives little information
on the dynamics of element turnover. It is known that levels of
elements may vary in tissues which are histologically different,
even within a single organ, e.g., kidney and bone, but several
studies have found regional differences within tissues that appear
grossly homogeneous, e.g., different lobes of the liver [33,53].
Many studies investigating possible involvement of trace elements
in disease processes look for a difference in total levels of an
element in a tissue in persons dying with the disease compared with
levels in a similar but healthy group. In general, such studies
have been inconclusive and it may be more profitable to look for an
alteration in distribution of the element rather than a change in
the total amount.

As well as knowing where an element is located, we need to
know in what form it exists in order to elucidate fully its biochemical functioning and to relate alterations to the changes observed
in a deficiency. For example, iron must be present in hemoglobin
as Fe(II) in order for that molecule to function properly, but is
stored in ferritin as Fe(III). The function of many copper-containing
enzymes depends on the relative ease with which it changes oxidation
states: $Cu^{2+} \rightleftharpoons Cu^{+}$. The strength with which a metal bonds covalently
to organic substrates can also influence its metabolism and biochem-
ical functioning [54].

There are severe limitations to investigating the finer points
of speciation and localization of trace elements. Both types of
studies require extensive sample separation by techniques such as
ultracentrifugation and ion-exchange chromatography. The more a
sample is manipulated, the greater the opportunities to introduce
contamination, particularly when working with very low concentrations
and very small sample sizes. Some newer instrumentation, such as
ion or proton microprobes, x-ray and mass spectrometric surface
techniques, and in vivo neutron activation may help to improve

specificity and eventually allow trace elements to be analyzed
without disrupting structural components. Presently, however,
this field remains one mainly for high-technology, specialist
trace element laboratories [55].

4. CONCLUDING REMARKS

Research in the field of trace element nutrition is by no means of
purely academic interest. Deficiencies are widespread in both
geographic and demographic terms. It has been estimated that 10-90%
of women and children in many areas suffer from iron deficiency [56],
making it possibly the most common of all nutritional deficiencies.
In spite of the syndromes of iron and iodine having been recognized
for over 100 years, neither are yet fully overcome. Chronic zinc
deficiency causing growth retardation may be widespread in certain
areas of the world, even among groups that are otherwise apparently
well nourished [57]. Low intakes of selenium may contribute to a
cardiomyopathy endemic over large areas of China [18,39]. Less than
optimum intakes of fluorine may contribute to poor dental health in
many population groups in many parts of the world. Although not
nearly as widespread on a population basis, deficiencies of copper
and chromium also occur. No widespread human deficiencies of other
trace elements have been discovered and it is probable that such may
not occur for the newer essential trace elements like nickel, vanadium, and silicon. What information we do have suggests that the
requirements for the newer essential trace elements, as for manganese,
are much lower than measured intakes so that individuals taking a
varied diet are unlikely to develop a deficiency. However, for these
elements, as well as for iron and zinc, we need to know more about
bioavailability and factors which may alter requirements.

Modern medical practice is making more and more extensive use
of synthetic diets including those for total parenteral nutrition.
To date, acute deficiency syndromes of zinc, copper, chromium, selenium, and molybdenum have been reported in patients receiving

unsupplemented total parenteral nutrition [24]. Poor nutritional
status of zinc, copper, and selenium has also been observed in
children on dietetic treatment for metabolic disorders [58]. In
these cases, selenium status was related to a low intake, but zinc
intake was not low and poor status appeared to be due to a low bioavailability of zinc from the formulas used. It is quite possible
we may see deficiencies of other trace elements occurring in such
situations. Extensive changes in the food supply or dietary habits
of a group may also contribute in future to the development of marginal trace element deficiencies, such as with the widespread advice
in many nutritional and medical circles to increase the intake of
plant fiber in the diet. Uneven and unnecessary use of dietary
supplements of some elements, or lack of understanding of the need
for supplements of other elements, e.g., iodine, fluorine, may also
contribute to the development of a deficiency or toxicity syndrome.

Research in trace element nutrition is ultimately directed
toward understanding the mode of action of an element and its metabolism in the body [7]. This knowledge then serves as a basis for the
determination of requirements and the assessment of the nutritional
status of that element in an individual or population group. Where
nutritional status is less than optimal, it may be necessary to
formulate an intervention program which ensures an ultimate public
health benefit, such as the successful use of iodine-supplemented
table salt to eradicate goiter in many areas [59] and fluoridation
of water supplies [60].

Much remains to be done. For most essential trace elements,
not only the newer ones, there are large gaps in our understanding
of their mode of action and behavior in the body, and even in such
basic information as levels in foods and the normal dietary intake.
Until we can be sure of our methodology and analytical techniques,
such gaps will remain.

REFERENCES

1. W. Mertz, *Science, 213,* 1332 (1981).

2. E. Underwood, *Trace Elements in Human and Animal Nutrition,* 4th ed., Academic Press, New York, 1977.

3. N. T. Davies, *Phil. Trans. Roy. Soc. Lond., B294,* 171 (1981).

4. W. Mertz, *Phil. Trans. Roy. Soc. Lond., B294,* 9 (1981).

5. J. C. Smith and K. Schwarz, *J. Nutr., 93,* 182 (1967).

6. K. Schwarz, *Fed. Proc., Fed. Am. Soc. Exp. Biol., 33,* 1748 (1974).

7. W. Mertz, in *Trace Element Analytical Chemistry in Medicine and Biology* (P. Brätter and P. Schramel, eds.), Walter de Gruyter, Berlin, 1980, p. 727ff.

8. A. Galdes and B. L. Vallee, *Met. Ions Biol. Syst., 15,* 1 (1983).

9. (a) H. H. Sandstead, in *Disorders of Mineral Metabolism,* Vol. 1, *Trace Minerals* (F. Bronner and J. W. Coburn, eds.), Academic Press, New York, 1981, p. 93. (b) H. J. M. Bowen, *Environmental Chemistry of the Elements,* Academic Press, New York, 1979.

10. P. W. Winston, in *Disorders of Mineral Metabolism,* Vol. 1, *Trace Minerals* (F. Bronner and J. W. Coburn, eds.), Academic Press, New York, 1981, p. 295ff.

11. H. E. Ganther, D. G. Hafeman, R. A. Lawrence, R. E. Serfass, and W. G. Hoekstra, in *Trace Elements in Human Health and Disease,* Vol. 2, *Essential and Toxic Elements* (A. S. Prasad and D. Oberleas, eds.), Academic Press, New York, 1976, p. 165ff.

12. W. R. Beisel, *Am. J. Clin. Nutr., 35,* 417 (1982).

13. H. A. Schroeder and A. P. Nason, *Clin. Chem.* (Winston-Salem, N. C.), *17,* 461 (1971).

14. National Research Council, *Recommended Dietary Allowances,* 9th ed., National Academy of Sciences, Washington, D.C., 1980.

15. National Research Council, *Geochemistry and the Environment,* Vol. 1, *The Relation of Selected Trace Elements to Health and Disease,* National Academy of Sciences, Washington, D.C., 1974.

16. A. S. Prasad, *Zinc in Human Nutrition,* CRC Press, Boca Raton, Florida, 1979.

17. C. E. Casey and K. M. Hambidge, in *Zinc in the Environment,* Part 2, *Health Effects* (J. O. Nriagu, ed.), John Wiley and Sons, New York, 1980, p. 1ff.

18. X. Chen, G. Yang, J. Chen, X. Chen, Z. Wen, and K. Ge, *Biol. Trace Elem. Res., 2,* 91 (1980).

19. A. Cordano, in *Zinc and Copper in Clinical Medicine* (K. M. Hambidge and B. L. Nichols, eds.), Spectrum, New York, 1978, p. 119ff.

20. E. A. Doisy, in *Trace Substances in Environmental Health,* VI (D. D. Hemphill, ed.), University of Missouri, Columbia, Missouri, 1972, p. 193ff.

21. E. R. Monsen, L. Hallberg, M. Layrisse, D. M. Hegsted, J. D. Cook, W. Mertz, and C. A. Finch, *Am. J. Clin. Nutr., 31,* 134 (1978).

22. J. Bowering, A. M. Sanchez, and M. I. Irwin, *J. Nutr., 106,* 985 (1976).

23. N. W. Solomons, *Am. J. Clin. Nutr., 35,* 1048 (1982).

24. P. A. Walravens, in *Human Nutrition, Clinical and Biochemical Aspects* (P. J. Garry, ed.), American Association for Clinical Chemistry, Washington, D.C., 1981, p. 398ff.

25. K. M. Hambidge and P. A. Walravens, *Clin. Gastroenterol., 11*(1), 87 (1982).

26. J. McC. Howell, J. M. Gawthorne, and C. L. White (eds.), *Trace Element Metabolism in Man and Animals, 4,* Australian Academy of Science, Canberra, 1981.

27. P. Brätter and P. Schramel (eds.), *Trace Element Analytical Chemistry in Medicine and Biology,* Walter de Gruyter, Berlin, 1980.

28. J. E. Spallholz, J. L. Martin, and H. E. Ganther (eds.), *Selenium in Biology and Medicine,* AVI, Westport, Connecticut, 1981.

29. L. Fowden, G. A. Ganton, and C. F. Mills (eds.), *Trace Element Deficiency: Metabolic and Physiological Consequences,* The Royal Society, London, 1981.

30. A. S. Prasad (ed.), *Clinical, Biochemical and Nutritional Aspects of Trace Elements,* Alan R. Liss, New York, 1982.

31. New Zealand Workshop on Trace Elements in New Zealand, University of Otago, Dunedin, New Zealand, 1981.

32. O. A. Levander and L. Cheng (eds.), Micronutrient Interactions: Vitamins, Minerals and Hazardous Elements, *Ann. N.Y. Acad. Sci., 355* (1980).

33. M. Milner (ed.), Research Needed to Improve Data on Mineral Content of Human Tissues, *Fed. Proc., Fed. Am. Soc. Exp. Biol., 40,* 2111 (1981).

34. International Atomic Energy Agency, *Elemental Analysis of Biological Materials. Current Problems and Techniques with Special Reference to Trace Elements,* Technical Report Series No. 197, IAEA, Vienna, 1980.

35. J. A. Rotruck, A. L. Pope, H. E. Ganther, and W. G. Hoekstra, *Science, 179,* 588 (1973).

36. J. Versieck and R. Cornelis, *Anal. Chim. Acta, 116,* 217 (1980).

37. R. M. Parr, in *Trace Element Analytical Chemistry in Medicine and Biology* (P. Brätter and P. Schramel, eds.), Walter de Gruyter, Berlin, 1980, p. 631ff.

38. C. F. Mills, *Chem. Br., 15,* 512 (1979).

39. A. T. Diplock, in *Selenium in Biology and Medicine* (J. E. Spallholz, J. L. Martin, and H. E. Ganther, eds.), AVI, Westport, Connecticut, 1981, p. 303ff.

40. I. Lombeck, K. Kasperek, L. E. Feinendegen, and H. J. Bremer, in *Selenium in Biology and Medicine* (J. E. Spallholz, J. L. Martin, and H. E. Ganther, eds.), AVI, Westport, Connecticut, 1981, p. 269ff.

41. C. D. Thomson and M. F. Robinson, *Am. J. Clin. Nutr., 33,* 303 (1980).

42. M. F. Robinson, in *Clinical, Biochemical and Nutritional Aspects of Trace Elements* (A. S. Prasad, ed.), Alan R. Liss, New York, 1982, 325ff.

43. C. F. Mills, *Phil. Trans. Roy. Soc. Lond., B294,* 199 (1981).

44. M. F. Robinson and C. D. Thomson, in *Selenium in Biology and Medicine* (J. E. Spallholz, J. L. Martin, and H. E. Ganther, eds.), AVI, Westport, Connecticut, 1981, p. 283ff.

45. C. F. Mills, *Fed. Proc., Fed. Am. Soc. Exp. Biol., 40,* 2138 (1981).

46. H. E. Ganther, *Ann. N.Y. Acad. Sci., 355,* 212 (1980).

47. C. H. Hill and G. Matrone, *Fed. Proc., Fed. Am. Soc. Ex. Biol., 29,* 1471 (1970).

48. V. R. Young and M. Janghorbani, *Cereal Chem., 58,* 12 (1981).

49. R. D. H. Stewart, N. M. Griffiths, C. D. Thomson, and M. F. Robinson, *Br. J. Nutr., 40,* 45 (1978).

50. M. F. Robinson and C. D. Thomson, *Nutr. Abstr. Rev., 53,* 3 (1983).

51. M. R. Spivey Fox, R. M. Jacobs, A. O. L. Jones, B. E. Fry, M. Rabowska, R. P. Hamilton, B. F. Harland, C. L. Stone, and S. H. Tao, *Cereal Chem., 58,* 6 (1980).

52. O. A. Levander, *Fed. Proc., Fed. Am. Soc. Exp. Biol., 42,* 1721 (1983).

53. G. V. Iyengar, W. E. Kollmer, and H. J. M. Bowen, *The Elemental Composition of Human Tissues and Body Fluids,* Verlag Chemie, Weinheim, 1978.

54. R. J. P. Williams, *Phil. Trans. Roy. Soc. Lond., B294,* 57 (1981).

55. E. Sabbioni, in *Trace Element Analytical Chemistry in Medicine and Biology* (P. Brätter and P. Schramel, eds.), Walter de Gruyter, Berlin, 1980, p. 407ff.

56. World Health Organization, *Control of Nutritional Anemia with Special Reference to Iron Deficiency,* Tech. Rep. Series No. 580, WHO, Geneva, 1975.

57. K. M. Hambidge, *Phil. Trans. Roy. Soc. Lond., B294,* 129 (1981).

58. I. Lombeck and H. J. Bremer, *Nutr. Metab., 21,* 49 (1977).

59. B. S. Hetzel and I. B. Hales, in *Endemic Cretinism* (J. B. Stanbury and B. S. Hetzel, eds.), John Wiley and Sons, New York, 1980, p. 123ff.

60. Royal College of Physicians, *Fluoride, Teeth and Health,* Pitman Medical, Turnbridge Wells, UK, 1976.

61. F. Bronner and J. W. Coburn (eds.), *Disorders of Mineral Metabolism,* Vol. 1, *Trace Minerals,* Academic Press, New York, 1981.

Chapter 2

METALS AND IMMUNITY

Lucy Treagan
Department of Biology
University of San Francisco
San Francisco, California

1. INTRODUCTION

In recent years a great deal of epidemiological and experimental
evidence has accumulated to show that various functions of the immune
system can be affected by exposure to metal compounds. This chapter
is intended as a summary rather than a comprehensive survey of some

of the more recent and current work in the area of metal interactions
with the immune system; earlier experimental and epidemiological
studies have been covered in previous reviews [1,2]. Possible mechanisms of the observed effects of metals on immune functions will be
explored and any correlation between metal carcinogenicity and the
nature of the effect of the metal on the immune system will be
evaluated.

1.1. Concepts of Immunization

The immune system of vertebrates is highly complex; its principal
function is to protect the host against infection by pathogenic
microorganisms and viruses as well as against tumor cells. A great
variety of organic molecules are capable of stimulating the immune
system and inducing an immune response; such molecules are known as
antigens. Most proteins, many polysaccharides, and other large
molecules are immunogenic, as are viruses, microorganisms, and
tissue cells recognized as foreign by the host. Small, chemically
defined molecules are not immunogenic by themselves but will stimulate the immune system if coupled with an immunogenic protein molecule. Such small molecules are known as haptens, while the protein
molecule linked to the hapten is a carrier. Stimulation of the
immune system may result in the synthesis of antibodies or in the
generation of specifically reactive lymphocytes. Antibody produced
in response to an antigen is capable of reacting with this antigen
(or hapten) in a highly specific manner. The nature of the reaction
is similar to that of enzyme and substrate. The antigen-antibody
complex is reversibly bound by noncovalent interactions, including
van der Waals forces, and electrostatic and hydrogen bonds. The
binding of antibody to a microbial antigen serves to neutralize
microbial activity and thus contributes to host defense.

1.2. Metals as Immunogens

A considerable amount of epidemiological and experimental evidence
points to immunogenicity of metals. Since small inorganic compounds
are not immunogenic, the actual metal immunogens are the covalent
derivatives formed by metals with tissue proteins in vivo; in case
of metal-induced skin hypersensitivity, the self-proteins of the
skin, specifically modified by covalent attachment of the haptenic
(metal) groups, are the principal antigenic units [3]. Such metal-
protein conjugates may interact with macrophages of the dermis; it
is postulated that the macrophages carrying the immunogenic complex
induce specific reactivity in certain subclasses of T lymphocytes.
Once such reactivity is established, the application of the sensi-
tizing metal to the skin will elicit a skin reaction that develops
in 6-8 hr and increases in size and intensity by 24-48 hr; such
reactions are characteristic of *delayed hypersensitivity* and are
mediated by specifically reactive T cells.

   The ability of metals to induce delayed hypersensitivity is
well documented from clinical and experimental studies [1,4]. Representative metals known to induce delayed hypersensitivity include
beryllium, chromium, cobalt, copper, gold, manganese, mercury, molybdenum, nickel, platinum, and zirconium. Hypersensitivity to chromium
is one of the more common clinical allergies and is often associated
with a reactivity to cobalt or nickel compounds [5,6]. In addition
to clinical studies of metal hypersensitivity, experimental studies
have demonstrated that the sensitization of guinea pigs with various
metals salts (beryllium, copper, molybdenum, chromium, cobalt, nickel,
manganese) is capable of causing skin reactivity of the delayed type
[7,8].

   The immunogenic capacity of metals includes the ability to
induce antibody. Platinum, nickel, and beryllium have been demonstrated in clinical and experimental studies to stimulate antibody-
mediated reactions [9-15].

Of particular interest are immune reactions induced by mercury and gold; mercury induces a high level of IgE in Brown Norway rats [16]; repeated injection of mercuric chloride into this strain of rats results in an immune complex type-glomerulonephritis [17].

Similarly, gold salts, used to treat patients with rheumatoid arthritis, induce immune complex type-glomerulonephritis in approximately 5% of treated patients [18].

The studies reviewed above provide evidence that metals have an extensive capacity for immunogenicity due, most probably, to their ability to form immunogenic complexes with tissue proteins.

2. EXPERIMENTAL STUDIES

2.1. Effect of Metals on Different Parameters of the Immune System

2.1.1. Phagocytosis

The polymorphonuclear leukocytes and the mononuclear phagocytes (monocytes and macrophages) constitute the first line of host defense against infectious agents. These cell types are responsible for the phagocytosis and intracellular killing of microbial pathogens. In addition to these functions, macrophages act as regulatory and accessory cells in the induction and expression of humoral and cellular immunity and under special circumstances are able to destroy tumor cells [19]. In view of these numerous and diverse functions, the nature of interaction of trace metals with macrophages is of major importance in understanding the effect of metals on the immune system.

A number of experimental and clinical studies have attempted to elucidate the effect of lithium on phagocytic cells. The activities of lymphocytes, macrophages, and polymorphonuclear leukocytes are regulated by alterations in cyclic nucleotide levels; lithium, as an adenylate cyclase inhibitor, has a profound effect on lymphocytes and on phagocytic cells. Lithium increases the number of polymorphonuclear leukocytes in circulation and is therefore given to patients with chemotherapy-induced neutropenia; lithium also enhances the

phagocytosis of latex particles by macrophages in direct proportion
to metal concentration [20].

Two other metals have an enhancing effect on macrophage
activity: aluminum and zirconium salts increase the phagocytic
activity of rabbit bronchopulmonary macrophages, as determined by
an in vitro phagocytic assay [21]. By contrast, the effect of gold
salts on phagocytic cells is generally inhibitory: macrophages incubated with gold salts in vitro are deficient in supporting mitogen-
induced T-cell proliferation; macrophage migration, chemotaxis, and
phagocytosis are also inhibited [22]. Incubation of human monocytes
with 25 µg/ml gold sodium thiomalate resulted in the appearance of
large, intracytoplasmic vacuoles and in a reduced phagocytic capacity
[23]. A similar depression in the phagocytic function was brought
about by cadmium [24]: the phagocytic capacity of alveolar and peritoneal macrophages, and of the polymorphonuclear leukocytes, was
diminished after in vitro incubation with cadmium chloride; alveolar
macrophages were particularly sensitive to cadmium effects and were
depressed in their intracellular microbicidal activity, while this
function was not affected in the peritoneal macrophages.

Additional studies of the effect of trace metals on phagocytic
cells are covered in a recent review [25].

2.1.2. Lymphocyte Activation

A number of substances of bacterial and plant origin are able to
stimulate lymphocytes to proliferate and differentiate into blast
cells. These mitosis-inducing agents or mitogens are used extensively in studies of the biochemistry of cell stimulation. Some
mitogens stimulate T cells whereas others interact with B cells, but
within either lymphocyte class many clones of cells will react with
the mitogen. Mitogens bind specifically to glycoprotein receptors
on the lymphocyte surface. Such binding initiates a series of intracellular events which include a pronounced development of rough endoplasmic reticulum and microtubules and an increase in the rates of
nucleic acid and protein synthesis. Plant lectins phytohemagglutinin

(PHA) and concanavalin (Con A) stimulate T cells; pokeweed mitogen stimulates both T and B cells. A mitogen of bacterial origin, *E. coli* lipopolysaccharide (LPS), acts on B cells causing them to secrete immunoglobulins which are predominantly of the IgM class [3].

Various metals affect the ability of lymphocytes to be activated by mitogens. Some of the earlier studies describing these effects have been reviewed [2]. Among recent experiments several studies focus on the effect of the lithium ion on lymphocyte response to mitogens [26-28]. The biochemical mechanism of action of this ion is of particular interest as it is widely used in psychiatry in the treatment of manic-depressive disease states [27]. Mononuclear cells from the peripheral blood of patients on lithium carbonate therapy show an enhanced blastogenic response when cultured in vitro in the presence of PHA and pokeweed mitogen. The observed increase in DNA synthesis may be the result of a direct effect of lithium on a sub-population of cells, stimulation of cells specifically responsive to PHA and pokeweed mitogen, or a modification of the activity of suppressor cells [28]. The addition of lithium chloride to cultures of mononuclear cells from the peripheral blood of healthy subjects enhanced lymphocyte proliferation in response to PHA; the increase in thymidine incorporation was parallel to lithium concentration [20, 29]. In parallel with the stimulation of human peripheral mononuclear cells, lithium chloride enhanced the response of hamster thymocytes to PHA and Con A stimulation [26]. Lithium effect on Con A-stimulated cultures depended on the dose of the mitogen; in the presence of suboptimal doses of Con A, lithium chloride enhanced thymidine incorporation; an opposite effect was noted with higher concentrations of Con A [27]. Other in vitro studies of metal effects on lymphocyte blastogenesis include the addition of gold, as sodium aurothiomalate, to cultures of rat lymphocytes stimulated with a bacterial mitogen prepared from *Mycoplasma pulmonis*: gold-induced inhibition of blastogenesis in proportion to metal concentration was observed [30]. Inhibition of lymphocyte activation was also demonstrated in other studies with gold: when human peripheral blood

mononuclear cells were incubated in vitro with pokeweed mitogen in
the presence of 10 μg/ml of gold thiomalate, an inhibition of lymphocyte differentiation into immunoglobulin-secreting cells was observed
[31]. Similarly, gold sodium thiomalate suppressed the response of
murine splenic lymphocytes to the B-cell mitogen LPS and to the T-cell
mitogen Con A [32]. Lymphocyte stimulation by T-cell mitogens is
enhanced in the presence of macrophages: gold-mediated inhibition of
T-cell-dependent reactions may be due to a primary interference with
monocytes or macrophages; macrophages incubated with gold salts are
deficient in supporting Con A-induced T-cell proliferation. Furthermore, untreated monocytes restore high levels of DNA synthesis to
gold-treated, mitogen-stimulated cultures of human peripheral blood
mononuclear cells [32]. In studies with other metals a diphasic,
dose-dependent effect on lymphocyte blastogenesis was observed: low
concentrations of manganese sulfate enhanced the mitogenic response
of hamster lymphocytes to Con A, PHA, and LPS; the addition of high
concentrations of manganese to cell cultures inhibited the response
to Con A and to PHA without any inhibitory effect on LPS-induced
mitogenesis [33]. Chloride salts of cadmium and lead either enhanced
or depressed the blastogenesis of mouse lymphocytes in response to
LPS, depending on metal concentration [34]. The capacity of lead to
stimulate the lymphocyte response to B-cell mitogens was confirmed
by Lawrence [35]: chloride salts of lead and nickel enhanced mitogenesis in mouse spleen cell cultures in response to B-cell mitogens;
zinc chloride was either inhibitory or stimulatory, depending on
metal concentration, while mercuric chloride was uniformly suppressive. The effect of zinc may be age-dependent: zinc chloride enhanced the blastogenic response of human peripheral blood lymphocytes
to PHA when cells from young persons were used but had an inhibitory
effect on lymphocytes from old individuals [36]. The effect on
lymphocyte blastogenesis of zinc, lead, cadmium, and mercury was
studied by Gaworski and Sharma [37]: cadmium and mercury inhibited
the response of mouse splenic lymphocytes to PHA and to pokeweed
mitogen, lead enhanced the blastogenic response, and zinc enhanced
the lymphocyte response to PHA but suppressed that to pokeweed

mitogen. When mice were exposed for 30 days to each of these metals
in drinking water, the mitogenic response of splenocytes to both
mitogens was significantly depressed. A similar inhibition of lymphocyte blastogenesis in response to PHA stimulation was reported after
oral administration of lead to mice and rats [38-40]. These and additional studies dealing with metal effects on lymphocyte blastogenesis
are summarized in Table 1. Some metals are able to stimulate lymphocyte blastogenesis independent of the presence of plant or bacterial
mitogens. The ability of zinc to induce the proliferation and division of resting lymphocytes is well documented [42]. The lymphocytes
that respond to zinc stimulation are predominantly T cells, although
a small proportion of B cells may also be involved. The in vitro
stimulation of human peripheral lymphocytes by zinc requires the
presence of monocytes, as shown by reduced mitogenic reactivity of
monocyte-depleted lymphocyte populations; the addition of cell culture
supernatants containing monocyte-derived factors substantially increases lymphocyte blastogenesis [42]. Mitogenic activity of lithium
has been demonstrated in cultures of hamster lymph node cells and,
to a lesser extent, splenocytes; when added to cultures of hamster
thymocytes, lithium failed to initiate blastogenesis [27]. Lead,
cadmium, chromium, and nickel are other metals with reported mitogenic activity [34,35]; earlier studies on metal-induced mitogenesis
have been covered in a previous review [2].

2.1.3. Antibody Synthesis

The presentation of an immunogen to a population of lymphocytes may
result in a specific interaction of the immunogen with receptor
molecules located on the plasma membrane of certain lymphocytes.
This interaction signals blastogenesis and differentiation of B
lymphocytes into antibody-secreting plasma cells. The majority of
known immunogens, including metal-protein conjugates, require the
participation of macrophages and a subclass of T cells known as
T-helper cells for effective B-cell stimulation. Such immunogens
are called T-dependent, in contrast to T-independent antigens that

TABLE 1

Effect of Metals on Stimulation of Lymphocytes by Mitogens

Metal	Mitogen	Metal effect on blastogenesis
Aluminum	PHA, LPS	Inhibition (metal-mediated induction of suppressor cells) [41]
Cadmium	LPS	Enhancement or inhibition, depending on metal concentration [34]
	PHA, pokeweed mitogen	Inhibition [37]
Gold	*Mycoplasma pulmonis*	Inhibition [30]
	Pokeweed mitogen	Inhibition [31]
	LPS, Con A	Inhibition [32]
Lead	PHA, pokeweed mitogen	Enhancement or inhibition, depending on experimental conditions [37]
	LPS, 2-mercaptoethanol	Enhancement [35]
	LPS	Enhancement or inhibition, depending on metal concentration [34]
	PHA	Inhibition [38-40]
Lithium	PHA	Enhancement [26]
	PHA, pokeweed mitogen	Enhancement [28]
	Con A	Enhancement or inhibition, depending on mitogen dose [27]
	PHA	Enhancement [29]
Manganese	Con A, PHA, LPS	Inhibition, enhancement, or no effect, depending on metal concentration [33]
Mercury	LPS, 2-mercaptoethanol	Inhibition [35]
	PHA, pokeweed mitogen	Inhibition [37]
Nickel	LPS, 2-mercaptoethanol	Enhancement [35]
Zinc	LPS, 2-mercaptoethanol	Enhancement or inhibition, depending on metal concentration [35]
	PHA	Enhancement or inhibition, depending on age of lymphocytes [36]
	PHA	Enhancement [37]
	Pokeweed mitogen	Inhibition [37]

can interact with and stimulate B cells without participation of
other cell types. The capacity of various metals to affect the
antibody response is well documented [1,2]. Most of the earlier
studies were conducted in vivo; these studies clearly demonstrated
a variable effect of metals on antibody synthesis, although chronic
exposure to metals by the oral route generally produced immunosuppression. The outcome of in vivo studies is profoundly affected by
a number of factors, including the type of antigen, animal species,
and hormonal influences [43-47]. Experiments with the immunological
adjuvant aluminum hydroxide produced either suppression or enhancement of the antibody response of mice to sheep red blood cells,
depending on the route of antigen administration [41]. Some additional in vivo experiments (Table 2) demonstrate metal-induced
immunosuppression [48-50]. Metal effects on antibody synthesis
have also been explored in vitro: Kutz et al. tested chloride salts
of cadmium, chromium, lead, mercury, and nickel for their effect on
the early phase of antibody synthesis [51]. In these experiments
the antibody response to sheep red blood cells by mouse splenocytes
was assayed by the hemolytic plaque technique. Chromium and lead
had no effect on antibody synthesis, while the remaining metals were
both cytotoxic and immunosuppressive. Lawrence [35] also used the
in vitro hemolytic plaque assay to test the immunomodulatory effects
of the chloride salts of mercury, copper, manganese, cobalt, cadmium,
chromium, tin, zinc, iron, calcium, lead, and nickel. Calcium and
iron had no effect on the antibody response of mouse splenocytes;
varying degrees of immunosuppression were produced by all remaining
metals with the exception of nickel and lead, which had an enhancing
effect on antibody synthesis (Table 3).

2.1.4. Cell-Mediated Reactions

The effect of metals on cell-mediated reactions have been studied
both in vivo and in vitro. The ability of mice and rats to develop
delayed hypersensitivity in response to antigenic stimulation was
suppressed after oral or intraperitoneal administration of cadmium

TABLE 2

Effect of Metals on Antibody Synthesis in vivo

Metals	Route of metal administration	Duration of metal treatment	Antibody response
Selenium	Oral	3-4 weeks	Suppression in male but not in female chickens [43]
Lead	Oral	12 weeks	No effect [44]
Lead	Oral	5-7 weeks	Suppression or no effect, depending on antigen [45]
Lead	i.p.	Single injection	Suppression [47]
Cadmium	Oral	10 weeks	No effect [46]
Cadmium	i.p.	Single injection	Suppression [47]
Aluminum	i.p.	Single injection	Enhancement or suppression, depending on the route of antigen administration [41]
Chromium	Metal salt added to freshwater	2 weeks	Suppression [49]
Lithium	i.p.	30 days	Suppression of primary response [50]
Zinc	s.c.	2 weeks	Suppression [48]
Nickel	s.c.	2 weeks	Suppression [48]
Zinc and nickel	s.c.	2 weeks	Suppression [48]

TABLE 3

Effect of Metals on Antibody Synthesis in vitro

Metal	Effect on antibody
Cadmium	Suppression [51]
Cadmium	Suppression [35]
Mercury	Suppression [51]
Mercury	Suppression [35]
Nickel	Suppression [51]
Nickel	Enhancement [35]
Chromium	No effect [51]
Chromium	Suppression [35]
Lead	No effect [51]
Lead	Enhancement [35]
Copper	Suppression [35]
Calcium	No effect [35]
Iron	No effect [35]
Manganese	Suppression [35]
Cobalt	Suppression [35]
Tin	Suppression [35]
Zinc	Suppression [35]

acetate, lead acetate, or lithium chloride [39,46,50]. The mixed lymphocyte reaction was used to study the effect of metals on cellular immunity in vitro: this reaction is based on the recognition of lymphocyte membrane antigens by another (responder) lymphocyte population. The addition of mercuric chloride to lymphocyte cultures inhibited cellular proliferation, while nickel and lead salts enhanced the response of lymphocytes to allogenic cells [35]. An inhibition of the mixed-lymphocyte reaction and cell-mediated cytotoxicity by gold sodium thiomalate was reported by Harth and Stiller [52]; gold also suppressed the in vitro generation of cytotoxic T cells [32].

TABLE 4

Effect of Metals on Cell-Mediated Immune Reactions

Metal	Type of immune reaction	Effect of metal
Lead	Delayed hypersensitivity	Inhibition [39]
Lead	Mixed lymphocyte reaction	Enhancement [35]
Cadmium	Delayed hypersensitivity	Inhibition [46]
Lithium	Delayed hypersensitivity	Inhibition [50]
Mercury	Mixed lymphocyte reaction	Inhibition [35]
Nickel	Mixed lymphocyte reaction	Enhancement [35]
Gold	Mixed lymphocyte reaction	Inhibition [52]
Gold	Cell-mediated cytotoxicity	Inhibition [32]
Strontium	Natural killer cell cytotoxicity	Inhibition [53]

Normal human peripheral blood contains a population of large lymphocytes with intracytoplasmic azurophilic granules. These cells are cytotoxic for tumor-derived or virus-infected cells and are called natural killer cells. Preincubation of these cells with strontium chloride suppressed their cytolytic activity [53]. The effects of metals on cell-mediated immune reactions are summarized in Table 4.

2.1.5. Interferon Synthesis

Interferons are glycoproteins whose biological effects include antiviral activity, regulation of the immune response, and inhibition of cell division. Relatively few studies have explored the effect of metals on interferon synthesis or activity. Among the metals tested, nickel was demonstrated to suppress interferon synthesis or antiviral action [54-56]. Metallic copper and aluminum also had a suppressive affect on interferon synthesis, with copper most active of the three metals tested [56]. Cadmium sulfate, used in near-toxic concentrations, inhibited the antiviral activity of interferon [55]; by contrast, cadmium chloride, when used in nontoxic concentrations, had

no effect on interferon synthesis [54]. Some metals can enhance
rather than suppress interferon synthesis or activity: in the course
of experiments on interferon purification, it was found that chloride
salts of a number of elements in the lanthanide series enhanced the
antiviral activity of both human and mouse interferons [57].

2.2. Possible Mechanisms

The induction of the immune response to most antigens involves the
participation of more than one cell type. Metals may exert their
modulatory effects on immunity by a direct interaction with any of
these cell types; such interaction may affect the cells at various
target sites. Membrane permeability changes have been demonstrated
following exposure to lead and mercury [58]. These changes may possibly be mediated by a direct effect of the metal on a membrane-bound
ATPase enzyme. The Na/K ATPase plays a role in the control of solute
transport across membranes; cadmium, beryllium, mercury, nickel, lead,
lithium, and zinc may modulate or inhibit the activity of the membrane
ATPase [58-60]. Changes in cell membrane permeability may also be
brought about through the disturbance of essential trace element
metabolism or through metal-induced peroxidation of membrane lipids
[58]. Zinc, in particular, has important functional effects on the
cell membrane of phagocytic cells: increased zinc concentration
inhibits membrane fluidity of macrophages and neutrophils, and decreases their bactericidal activity [25]. Additional cellular targets for metal activity include effects on protein, RNA, and DNA
synthesis [58]. Cadmium, lead, arsenic, and methylmercury affect
tissue protein synthesis; RNA synthesis may be enhanced by methylmercury, nickel inhibits RNA polymerase activity, while the effect
of cadmium on RNA synthesis is dose-dependent. A number of metals
alter the fidelity of DNA synthesis; these include silver, beryllium,
cadmium, cobalt, copper, manganese, nickel, and lead. Both mercury
and zinc bind to cellular DNA; such binding may explain the ability
of these metals to stimulate lymphocyte DNA synthesis [61].

2.3. Metal-Metal Interactions

Mutual interaction between two metals can be either antagonistic or
synergistic. Metal interaction in vivo is affected by variables
such as metal concentration, length of exposure, and age of experimental animals. Treatment of rats with zinc plus nickel further
depressed their antibody response and decreased the number of blast
cells in spleens [48]. The in vitro antibody response of mouse
splenocytes was enhanced by nickel and lead; the combined use of
these metals produced similar effects [35]. Lithium and magnesium
had a synergistic effect on lymphocyte stimulation while the suppression by lithium of Con A-mediated blastogenesis was reversed
by potassium; the latter effect may be due to the effect of lithium
on Na/K ATPases [27].

2.4. Metal Carcinogenesis and the
Effect of Metals on Immunity

The effects on the immune system of the known metal carcinogens
chromium, lead, cadmium, and nickel range from enhancement to
suppression (Tables 1-3). The experimental data reviewed in this
chapter do not show any apparent correlation between metal carcinogenicity and a specific type of effect of the metal on the immune
system.

3. CONCLUSIONS

Metals affect various types of immune reactions. These include the
stimulation of lymphocytes by mitogens or antigens, antibody synthesis, T-cell-mediated reactions, and natural killer cell activity.
The mechanisms of metal interaction with the immune system are not
known; they may include metal effect on cell membrane permeability
and on cellular DNA, RNA, and protein synthesis. The treatment of

animals or lymphocyte cultures with metals frequently results in
immunosuppression, but an enhancement of the immune response may
also occur. It therefore appears that metal effects on immuno-
competent cells may be immunoregulatory rather than merely immunosuppressive.

ABBREVIATIONS AND DEFINITIONS

Abbreviations

Con A	Concanavalin A	
Ig	Immunoglobulin	IgG, IgM, IgA, IgD, and IgE are different Ig classes
i.p.	Intraperitoneal	
LPS	Lipopolysaccharide	
PHA	Phytohemagglutinin	
PMN	Polymorphonuclear leukocytes	
s.c.	Subcutaneous	

Definitions

Adjuvant	any foreign material introduced with an antigen to enhance its immunogenicity
Allogeneic	originating from a different individual of the same species
Antibody	an immunoglobulin capable of specific binding to an antigen
Antigen	a molecule capable of stimulating an immune response
B cell	one of the two major classes of lymphocytes. B cells respond to antigen by differentiating into antibody-producing cells
Blast cell	a cell undergoing the process of division

Blastogenesis   the production of blast cells, generally in
                       association with activation of cells by antigen
                       or mitogen

Humoral immunity   immune reactions involving the production of
                       specific antibody

Immunogen   a molecule capable of inducing an immune response

Immunoglobulin   gammaglobulin molecules having antibody activity

T cell   a class of lymphocytes that differentiate in the
                       thymus and mediate cellular immunity

REFERENCES

1. L. Treagan, *Res. Commun. Chem. Pathol. Pharmacol.*, *12*, 189
   (1975).

2. L. Treagan, *Biol. Trace Elem. Res.*, *1*, 141 (1979).

3. B. D. Davis, R. Dulbecco, H. N. Eisen, and H. S. Ginsberg,
   Microbiology, Harper and Row, New York, 1980, pp. 394-395, 505.

4. G. Kazantzis, *Environ. Health Persp.*, *25*, 111 (1978).

5. J. Thormann, N. B. Jespersen, and H. D. Joensen, *Contact Dermatitis*, *5*, 261 (1979).

6. I. Rystedt, *Contact Dermatitis*, *5*, 233 (1979).

7. O. B. Karpenko, *Gig. Tr. Prof. Zabol.*, *2*, 53 (1979).

8. A. Boman, J. E. Wahlberg, and G. Hagelthorn, *Contact Dermatitis*,
   5, 332 (1979).

9. O. Cromwell, J. Pepys, W. E. Parish, and E. G. Hughes, *Clin.
   Allergy*, *9* (1979).

10. J. Pepys, W. E. Parish, O. Cromwell, and E. G. Hughes, *Monogr.
    Allergy*, *14* (1979).

11. V. A. Tomilets, V. I. Dontsov, and I. A. Zakharova, *Zh. Mikro-
    biol. Epidemiol. Immunobiol.*, *8*, 126 (1978).

12. J. L. Malo, A. Cartier, M. Doepner, E. Nieboer, S. Evans, and
    J. Dolovich, *J. Allergy Clin. Immunol.*, *69*, 55 (1982).

13. N. K. Veien, A. H. Christiansen, E. Svejgaard, and K. Kaaber,
    Contact Dermatitis, *5*, 378 (1978).

14. V. Bencko, E. V. Vasilieva, and K. Simon, *Environ. Res.*, *22*,
    439 (1980).

15. M. J. Cianciara, A. P. Volkova, N. L. Aizina, and O. G. Alekseeva, *Int. Arch. Occup. Environ. Health, 45,* 87 (1980).

16. A. Prouvost-Danon, A. Abadie, C. Sapin, H. Bazin, and P. Druet, *J. Immunol., 126,* 699 (1981).

17. J. F. Bernaudin, E. Druet, M. F. Belair, M. C. Pinchon, C. Sapin, and P. Druet, *Clin. Exp. Immunol., 38,* 265 (1979).

18. B. Skrifuars, *Scand. J. Rheumatol., 8,* 113 (1979).

19. D. S. Nelson, *Clin. Exp. Immunol., 45,* 225 (1981).

20. L. Shenkman, W. Borkowsky, and B. Shopsin, *Med. Hypoth., 6,* 1 (1980).

21. R. P. Stankus, M. R. Schuyler, R. A. D'Amato, and J. E. Salvaggio, *Infect. Immun., 20,* 847 (1978).

22. M. Harth, *J. Rheumatol., 6,* Suppl. 5, 7 (1979).

23. P. E. Lipsky, K. Ugai, and M. Ziff, *J. Rheumatol., 6,* Suppl. 5, 130 (1979).

24. L. D. Loose, J. B. Silkworth, and D. W. Simpson, *Infect. Immun., 22,* 378 (1978).

25. W. R. Beisel, *Am. J. Clin. Nutr., 35,* Suppl., 417 (1982).

26. D. A. Hart, *Cell. Immunol., 58,* 372 (1981).

27. D. A. Hart, *Cell. Immunol., 43,* 113 (1979).

28. L. A. Fernandez and R. A. Fox, *Clin. Exp. Immunol., 41,* 527 (1980).

29. L. Shenkman, W. Borkowsky, R. S. Holzman, and B. Shopsin, *Clin. Immun. Immunopathol., 10,* 187 (1978).

30. Y. Naot and S. Merchav, *Infect. Immunol., 21,* 340 (1978).

31. S. A. Rosenberg and P. E. Lipsky, *J. Rheumatol., 6,* Suppl. 5, 107 (1979).

32. J. J. Jennings, S. Macrae, and R. M. Gorczynski, *Clin. Exp. Immunol., 36,* 260 (1979).

33. D. A. Hart, *Exp. Cell. Res., 113,* 139 (1978).

34. K. Gallagher, W. Matrazzo, and I. Gray, *Fed. Proc., 37,* 377 (1978).

35. D. A. Lawrence, *Toxicol. Appl. Pharmacol., 57,* 439 (1981).

36. K. M. K. Rao, S. A. Schwartz, and R. A. Good, *Cell. Immunol., 42,* 270 (1979).

37. C. L. Gaworski and R. P. Sharma, *Toxicol. Appl. Pharmacol., 46,* 305 (1978).

38. B. A. Neilan and L. Taddeini, *Clin. Res., 27,* 223A (1979).

39. R. E. Faith, M. I. Luster, and C. A. Kimmel, *Clin. Exp. Immunol.*, *35*, 413 (1979).

40. B. A. Neilan, L. Taddeini, C. E. McJilton, and B. S. Handwerger, *Clin. Exp. Immunol.*, *39*, 746 (1980).

41. N. Hanna, S. Blanc, and D. Nelken, *Cell. Immunol.*, *53*, 225 (1980).

42. H. Rühl and H. Kirchner, *Clin. Exp. Immunol.*, *32*, 484 (1978).

43. J. A. Marsh, R. R. Dietert, and G. F. Combs, Jr., *Soc. Exper. Biol. Med. Proc.*, *166*, 228 (1981).

44. L. J. Hoffman, W. B. Buck, and L. Y. Quinn, *Am. J. Vet. Res.*, *41*, 331 (1980).

45. M. I. Luster, R. E. Faith, and C. A. Kimmel, *J. Environ. Pathol. Toxicol.*, *1*, 397 (1978).

46. S. Muller, K. E. Gillert, C. Krause, G. Jautzke, U. Gross, and T. Diamantstein, *Experientia*, *35*, 909 (1979).

47. J. G. O'Neill, *Bull. Environ. Contam. Toxicol.*, *27*, 42 (1981).

48. S. Mehrmofakham and L. Treagan, *Biol. Trace Elem. Res.*, *3*, 7 (1981).

49. R. H. Sugatt, *Arch. Environ. Contam. Toxicol.*, *9*, 207 (1980).

50. B. D. Jankovic, L. Popeskovic, and K. Isakovic, *Adv. Exp. Med. Biol.*, *114*, 339 (1979).

51. S. A. Kutz, R. D. Hindsill, and D. J. Weltman, *Environ. Res.*, *22*, 368 (1980).

52. M. Harth and C. R. Stiller, *J. Rheumatol.*, *6*, Suppl. 5, 103 (1979).

53. P. A. Neighbour and H. S. Huberman, *J. Immunol.*, *128*, 1236 (1982).

54. D. Pribyl and L. Treagan, *Acta Virologica*, *21*, 507 (1977).

55. J. H. Gainer, *Am. J. Vet. Res.*, *38*, 863 (1977).

56. N. Hahon, J. H. Booth, and D. J. Pearson, Proc. Meet. In Vitro Eff. Miner. Dusts (R. C. Brown, M. Chamberlain, and R. Davies, ed.), 1980, p. 219.

57. J. J. Sedmark and S. E. Grossberg, *J. Gen. Virol.*, *52*, 195 (1981).

58. K. S. Squibb and B. A. Fowler, *Environ. Health Persp.*, *40*, 181 (1981).

59. J. D. Robinson, *Biochem. Biophys. Acta*, *413*, 459 (1975).

60. M. Chvapil, *Life Sci.*, *13*, 1041 (1973).

61. H. Rühl, H. Kirchner, and G. Bochert, *Soc. Exp. Biol. Med. Proc.*, *137*, 1089 (1971).

Chapter 3

THERAPEUTIC CHELATING AGENTS

Mark M. Jones
Department of Chemistry and
Center in Environmental Toxicology
Vanderbilt University
Nashville, Tennessee

1. INTRODUCTION

1.1. Mode of Action of Therapeutic Chelating Agents

Metal ions are toxic because they can form bonds with enzymes and
other molecules in biological systems which prevent these molecules
from functioning in a normal manner. Therapeutic chelating agents
are compounds used as antidotes for metal poisoning. The mode of
action of these compounds is based on the fact that they can compete
for metal ions with binding sites on molecules in biological systems.
The reaction of the therapeutic chelating agent with the metal bound
to a biological site must generally be a reasonably rapid one. The
resultant complex may be a binary or ternary one, involving the metal
ion, the therapeutic chelating agent and possibly another ligand, an
amino acid or other molecule, and is generally water-soluble. The
usual route of excretion is via the bloodstream to the kidney, filtration out of the kidney, followed by excretion in the urine, where
the presence of an enhanced concentration of the metal may be confirmed by analysis. An alternative route of considerable importance

with some metals (such as copper and iron) is via the liver to the
bile with subsequent excretion in the feces. In acute metal poisoning, time is of the essence, and immediate treatment with a chelating
agent is necessary to minimize permanent damage to the individual.
During this critical period, the complexes of toxic heavy metals with
therapeutic chelating agents may damage the kidney when the *rate* at
which they are transported through the kidneys is too high. The
common consequence of such damage is greatly diminished urine volume
(oliguria) or no urine production at all (anuria). In such cases
hemodialysis or peritoneal dialysis allows the removal of the metal
complex to continue along with the removal of other waste products.
This type of damage usually clears up in 3-7 days and the excretion
of the toxic metal complex may even increase significantly with the
return of normal kidney function. This emphasizes the fact that the
toxicity of the metal chelates themselves needs to be considered
carefully in any therapeutic scheme.

1.2. The Theory of Therapeutic Chelate Action

There are many advantages to be gained from the development of an
effective model which describes the functioning of therapeutic
chelating agents in quantitative terms. The first useful model was
that developed by Schubert [1-4], who assumed that the critical
interaction took place in the serum and that this consisted of the
competition between serum calcium and the foreign metal ion for the
binding sites of the chelating agent Y:

$$CaY + M \rightleftharpoons MY + Ca^{2+}$$

for which

$$K = \frac{K_{MY}}{K_{CaY}}$$

i.e., the ratio of the stability constants. Schubert then calculated the ratio $[MY]/[M] = R_{MV}$ as

$$R_{MV} = \left(\frac{K_{MY}}{K_{CaY}}\right)\left(\frac{[CaY]}{[Ca]}\right)$$

where [Ca] is the concentration of calcium in the serum. The key
factor here is the ratio of the stability constant values for the
toxic metal and calcium. This equation, which was designed for use
with chelating agents of the EDTA type (which form very stable
calcium complexes) works reasonably well as a predictive tool for
studies of the efficacy of such chelating agents in removing lanthanides, actinides, and foreign ions with inert gas type of electronic
configurations. This approach with its emphasis on the competition
between the foreign ion and calcium breaks down when "soft" metal
ions, such as mercury, are considered, which generally bind to
different types of donor atoms than the calcium ion does, i.e., to
-SH.

Schubert's theory was subsequently elaborated by Catsch and
his co-workers [5,6] so that a number of complicating features could
be taken into consideration, such as the hydrolysis of the metal ion
and the presence of a variety of metal chelate species for a given
ion, by using the effective stability constant. Catsch also carried
out detailed experimental studies which refined the theory to the
point where it was quite successful in describing how the foreign
ion concentration of an organ varied during the course of treatment
with a chelating agent.

In recent years there have been a number of studies in which
computer programs have been developed to describe how a very large
number of competitive equilibria affect the distribution of a metal
ion among various natural and added chelating species in solution
[7-13]. At the present time they are strictly limited to consideration of equilibria in the serum, and in many cases these equilibria
do not play a prime role in the processes by which the toxic metal
is removed from the body. It is generally assumed that charged
chelating anions cannot penetrate the cell membranes at all, though
there seem to be situations where this appears to occur to some
extent. As a consequence the development of an effective computer

simulation model may require the use of a multicompartment model
of the sort used by pharmacologists. As yet, no computer simulation model involving the features of the stability constant model
plus those of the multicompartment model has been developed to
describe the operation of therapeutic chelating agents. This
would appear to be an area in which very rapid developments might
follow upon the development of a more effective computer simulation model.

1.3. Synergistic and Mixed Ligand Chelate Therapy

The fact that mixed complexes (with a single metal and two types of
ligands) may have higher stability constants than those with a single
chelating agent led Schubert to propose the use of such mixtures to
achieve an enhancement in the rates at which a toxic metal ion could
be complexed and removed from the mammalian body [14]. While some
data used to support this claim [15] have been called into question
[16-22], the fundamental idea is still theoretically valid and there
are also other reasons why mixtures of chelating agents or chelating
agents and other compounds may result in a significant enhancement
in the rate of excretion of a toxic metal ion [23]. A small but
growing body of experimental information [24,25] proves that mixtures
of compounds can be significantly more effective than their individual
components in enhancing the excretion of certain metal ions, specifically, iron, cadmium, lead, mercury, and plutonium. Where this involves two chelating agents, it may be called mixed ligand chelate
therapy; where it involves two compounds with quite different functions, synergistic chelate therapy is the more accurate descriptive
term [23].

1.3.1. Mixed Ligand Chelate Therapy

Mixtures of chelating agents have been shown by Volf [24] to be
significantly more effective than individual compounds in removing
freshly injected plutonium. The mixtures with this action included

diethylene triaminepentaacetic acid (DTPA) + dipicolinate, DTPA +
desferrioxamine (DFOA), DFOA + citrate, and DTPA + citrate. Cherian
[26] has shown that mixtures of BAL + DTPA are more effective than
either alone in decreasing the body burden of cadmium.

1.3.2. *Synergistic Chelate Therapy*

Here two compounds are used which have quite different modes of
action. Thus maleate increases the excretion of mercury due to
D-penicillamine [27], ascorbic acid increases the excretion of iron
due to desferrioxamine [28], spironolactone increases the excretion
of mercury due to 2,3-dimercaptopropane-1-sulfonate [29], and
ascorbic acid increases the excretion of lead caused by EDTA [30].

2. BRITISH ANTI-LEWISITE (2,3-DIMERCAPTOPROPANOL-1) AND OTHER DITHIOLS

The interaction of toxic heavy metals with biological systems frequently involves their bonding to the sulfhydryl groups of enzymes
as the sites most sensitive to their action. It was established
quite early that a temporary reversal of the toxicity due to such
a heavy metal [specifically arsenic(III)] could be obtained by the
administration of compounds which themselves contained sulfhydryl
groups and could thus compete with the biological binding sites for
the toxic species. Later, Sir Rudolph Peters [31,32] and a research
group searching for antidotes for the war gas Lewisite:

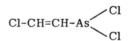

found that its action on pyruvate oxidase could be reversed by
various -SH-containing compounds. A very large number of compounds
containing one, two, and more -SH groups were then synthesized and
evaluated successively on the basis of (1) their ability to reverse
the action of arsenic on the enzyme pyruvate oxidase, (2) their own
inherent toxicity, and (3) their efficacy as antidotes (in animals)

to topically applied Lewisite and other arsenic compounds. Finally, because it was suspected that large amounts of the antidote might be required, the compound ultimately selected had to be one whose synthesis on a much larger than laboratory scale was feasible. In the course of these studies a number of vicinal dithiols were prepared and tested as well as other thiol compounds. The compound selected from this screening program was 2,3-dimercaptopropanol-1 or BAL (British Anti-Lewisite):

$$CH_2 - CH - CH_2$$
$$\ \ |\qquad\ |\qquad\ |$$
$$\ \ SH\quad SH\quad OH$$

BAL

This was the first chelating agent with clearly demonstrated therapeutic abilities to be used successfully for this purpose in humans.

2.1. British Anti-Lewisite (2,3-Dimercaptopropanol-1) (BAL)

BAL was first shown to be an effective antidote for arsenic(III) following the demonstration that it can remove arsenic(III) from an enzyme to which it is bound via mercaptan groups [33]. The arsenic(III)-BAL complex which is formed:

is sufficiently water-soluble to be excreted in the urine. Subsequent to its characterization as an antidote for arsenic(III) intoxication, it was tested as an antidote for most of the toxic heavy metals which form insoluble sulfides. It was found to be useful as an antidote for a number of such species, specifically: Pb^{2+}, Hg^{2+}, Sb(III), Au(I), and Cu(II) [34,35]. In most of these cases the toxic metal had been ingested in relatively modest amounts and the administration of BAL enhanced the urinary or fecal excretion of

the metal. Relatively few of the solid complexes of BAL with metal
ions have been characterized. The complexes with Hg^{2+} have been,
and it is found that Hg^{2+} forms two bonds at 180° with the S donors,
i.e., it is *not* a metal chelate; in this case the structure of the
1:1 complex is that of a polymer [36]. Because BAL was the first
therapeutically effective chelating agent available broadly and
approved for use in human patients, it has persisted in clinical use
after more effective and less toxic compounds have become available
as replacements. Because BAL is soluble in lipids it can penetrate
cell membranes much more readily than the usual ionic chelating
agents. The advantages which this confer on BAL include the ability
to mobilize deposits of toxic heavy metals not readily available to
attack by other chelating agents. The disadvantages stem from the
fact that the complexes formed may also possess sufficient lipid
solubility to cause problems. Thus, treatment with BAL can actually
increase the amount of Hg^{2+}, CH_3Hg^+, and $C_6H_5Hg^+$ transported into
the brain [37-40].

2.2. Unithiol (Sodium 2,3-Dimercaptopropane-1-sulfonate) (DMPS)

This compound:

$$CH_2 - CH - CH_2SO_3Na$$
$$\quad | \quad\quad |$$
$$\quad SH \quad\quad SH$$

Unithiol or DMPS

was among the very large number of mono- and dithiols first prepared
by Owen and his co-workers [41] and reported about 10 years after
BAL. It is more difficult to prepare than BAL and appreciably more
expensive, but in clinical use it has an enormous number of advantages over BAL. It is water-soluble, has a very low inherent toxicity, has a pattern of metal ion complexation very, very similar to
that of BAL, and it can be administered orally or by injection as
an aqueous solution in isotonic saline. Shortly after the preparation was reported by Owen and Johary, in 1955, Petrunkin [42]

published a slightly different preparation. The compound was subsequently prepared on a larger scale in the Soviet Union and the
majority of the literature on its applications has emerged from the
Soviet Union. The LD_{50} values for various modes of administration
of this compound in various species are generally of the order of
1000 mg/kg.

Unithiol (DMPS) forms very stable, water-soluble complexes
with Hg^{2+}, Pb^{2+}, Cd^{2+}, Zn^{2+}, Bi^{3+}, As^{3+}, Sb^{2+}, Sn^{2+}, and Ni^{2+}, among
others [43]. With Cu^{2+} a reduction occurs very rapidly which gives
a water-soluble Cu^+ complex. Polymeric complexes are also formed,
analogous to those of BAL, with nickel(II) and some other species.
The stability constants reported for the complexes of unithiol and
Hg^{2+} are very high, $\sim 10^{35}$ [44], and not surprisingly this compound
is a very effective antidote for acute mercuric chloride intoxication
[45].

Its administration to an animal which has a toxic heavy metal
distributed throughout its organs leads usually to the formation and
excretion of a stable water-soluble complex. After the administration of the compound, it moves from the site of injection or absorption quite rapidly, going to other parts of the body in the extracellular fluids. It then reacts with the metal, transforming it
into a water-soluble complex. This enters the serum and is then
transported to the kidney, which excretes it via the urine. In the
case of unithiol, the passage of the compound itself through the
body is very rapid, the half-life being about 2 hr or so.

The complexes of unithiol appear to be almost all very considerably *less* toxic than the metals from which they are derived,
an important factor in the wide effectiveness of this compound as
an antidote for toxic metals which bond readily to mercaptan groups
(generally that group which forms water-insoluble sulfides).

The manufacture of this compound by E. Heyl & Co., West Berlin,
has made it available in the West recently and a broader knowledge
of its many interesting properties should lead to its more extensive
clinical usage.

2.3.   2,3-Dimercaptosuccinic Acid (DMSA)

This compound:

$$HOOC - \underset{\underset{SH}{|}}{CH_2} - \underset{\underset{SH}{|}}{CH_2} - COOH$$

2,3 – Dimercaptosuccinic Acid (DMSA)

is another one of the vicinal dithiols first reported from Owen's
laboratory [46].   Its coordination chemistry [47-49] is essentially
the same as that of BAL and unithiol.   It forms very stable complexes
(usually water-soluble) with those ions which give insoluble sulfides.
It is more soluble in water than BAL and has a very low inherent
toxicity, being much less toxic than BAL (the oral LD_{50} value in
mice is reported as 5000 mg/kg [50]).   Its therapeutic applications
as a toxic metal antidote cover essentially the same metals as BAL
and unithiol, though the literature available on this compound is
not as extensive.   It can be given orally and is rapidly excreted
from the human body [51], is easier to prepare than unithiol, costs
about one-quarter as much, and is among the least toxic of all the
therapeutic chelating agents, though its clinical use in this role
is not common.   In the form of its antimony complex, it has been
used in the treatment of schistosomiasis [52].   While this antimony
complex is not as effective in killing the parasites as the potassium
antimonyl tartrate commonly used for this purpose, its use is accompanied by far fewer side effects than are found with potassium antimonyl tartrate [53].

Dimercaptosuccinic acid has been used with human patients with
lead and mercury poisoning [54].   For lead poisoning it is as effective as CaEDTA.   For mercury it seems also to be quite effective and
is also capable of enhancing the rate of excretion of methylmercury
from the brain [55].   In this respect it is quite superior to BAL,
which can increase the content of methylmercury in the brain.   In
fact this compound may well be the best compound presently available
for clearing methylmercury from the mammalian brain [56].   It can

also function as an antidote for arsenic(III) intoxication [57] and
in this respect is very superior to BAL or D-penicillamine, though
perhaps not quite so good as unithiol.

Because it can form very stable, water-soluble complexes with
the radioactive ions of technetium, it is employed, in the form of
such compounds, as a constituent of radiocontrast agents [58]. These
are compounds whose paths through the kidney can be followed by virtue
of their radioactivity. From the irregular nature of such paths in
pathological conditions, considerable information is furnished prior
to an operation.

2.4. Other Dithiols

Ever since the initial animal studies with BAL it was realized that
its applications would be limited by its own inherent toxicity. It
was also realized that this disadvantage was related to its limited
water solubility and fair lipid solubility. One consequence of this
lipid solubility was that BAL and some of its metal complexes were
sufficiently soluble in the lipids of the body that they were able
to penetrate the blood-brain barrier easily. This led to a search
for water-soluble derivatives which both possessed a lower level of
inherent toxicity and lacked the ability to pass through cell membranes as readily as BAL itself. Of the derivatives of BAL which
were synthesized, BAL glucoside had one of the most promising sets
of properties. The structure of this compound is

BAL–Glucoside

It is made by the reaction of glucose with allyl alcohol to give the
allylglucoside. This is transformed successively to the tetraacetyl

allyl glucoside, the tetraacetyl dibromopropyl glucoside, and the BAL glucoside [59].

The intravenous LD_{50} of BAL glucoside in the rabbit is greater than 5000 mg/kg [59]. It has been relatively little studied in comparison with BAL itself, though it is known to be an effective antidote for arsenic [59], lead [60], and cadmium [61]. At the present the main drawback in its use is the fact that it is not commercially available.

Another derivative of BAL, though also not available commercially, is N-(2,3-dimercaptopropyl)phthalamidic acid:

This has been reported to be an antidote for lead and mercury poisoning [62-64]. The compound is very effective in increasing the excretion of mercury and methylmercury in the bile and hence the feces, as well as increasing the urinary excretion of mercury. It is superior to BAL and to D-penicillamine in removing mercuric chloride from the tissues of animals [64].

Following the synthesis of sodium 2,3-dimercaptopropane-1-sulfonate, Petrunkin prepared a large number of structurally related dithiols which contained water-solubilizing sulfonate groups [65,66]. Some of these compounds bound the toxic heavy metals more firmly than unithiol itself, but little or no animal or clinical studies have been carried out on them as yet.

Although BAL was selected for development because it is an effective antidote for arsenic intoxication, it is *not* necessarily the compound that forms the most stable complexes with all the toxic heavy metals. Thus 1,3-dimercaptopropanol-2 is able to reverse the inactivation of pyruvate oxidase by cadmium at lower concentrations than BAL [67]. The whole question of the optimum "bite size" of dithiol chelating agents has not been investigated in any detail. From the information presently available on copper complexes with nitrogen donors, a mixture of six- and five-membered chelate rings can form complexes of appreciably greater stability than those in which only five-membered rings are possible. These ideas have not really been put to any test with polythiol chelating agents. It is thus possible that new members of this class with three or four suitably located thiol groups which possess appropriate groups to give the chelating agents a sufficient degree of water solubility will be found to have properties quite superior, for certain applications, to any members of this class presently known.

3. THE AMINOPOLYCARBOXYLIC ACIDS

These compounds may conveniently be thought to be that class of structures formed from ammonia and other amines by the successive replacement of N-H linkages by N-CH$_2$COOH structures. Literally hundreds of these compounds have been prepared and the literature on them is huge. Although a great variety of such structures have been prepared, the compounds do possess certain common features which simplifies their discussion. The first of these is a nearly universal basic arrangement of oxygen and nitrogen donor groups:

$$\begin{array}{c} \quad\quad | \quad\;\; | \quad\quad\quad \ddot{O}\!: \\ \_N\!-\!C\!-\!C \diagup \\ \diagup \; \ddot{}\; \;\; | \quad\quad \diagdown \; .\ddot{O}\!: \end{array}$$

which gives a five-membered metal chelate ring when both N and O are bonded to the same metal ion. This common pattern of donor groups (usually repeated several times) gives these compounds the ability to form stable chelate rings with most of the common metal ions. For ions with an inert gas electronic configuration the resulting metal chelates are often quite stable and useful as therapeutic chelating agents. As one moves across the periodic table toward softer ions and ions which do not have an inert gas type of electronic configuration, the complexes formed, while still quite stable, are not necessarily among the most stable ones. As a result these compounds and compounds with S donors must both be considered in searching for therapeutic chelating agents for these metal ions. When one looks at ions such as Hg^{2+}, one finds that although the complex with, say, EDTA is quite stable, it is not stable enough to pull Hg^{2+} ions from the S donor sites to which they bind in biological structures. As a consequence there are toxic soft ions for which EDTA is of no therapeutic interest, even though the complexes formed have high values of K_{stab}. The types of reactions which this type of compound undergoes with metal ions may be seen in the surveys of such reactions with EDTA [68].

3.1. EDTA

Because EDTA and other chelating agents of this class form very stable complexes with the calcium ions of the serum, the direct injection of such chelating agents in a saline solution buffered to the blood pH leads to an abrupt decrease in the free calcium ion concentration of the serum and the onset of a tetany which can be fatal [69]. As a consequence it is customary that chelating agents of this class are administered intravenously as their calcium complexes [70]. For some situations the zinc(II) complexes are administered, since the zinc complexes are often compounds of lower toxicity and greater stability than the calcium ones, their administration does not lead to a sudden depletion in the free calcium ion concentration of the serum.

Na$_2$CaEDTA was first used as an antidote for lead poisoning [70-72] and was very effective in removing lead by enhancing its urinary excretion. The distribution of freshly injected lead in the mammalian body is time-dependent [73]. In untreated lead poisoning the lead is slowly removed from the soft tissues by two separate processes: excretion (via the urine and feces) and deposition in the bone where the Pb^{2+} ions can occupy sites otherwise occupied by Ca^{2+} ions. The use of water-soluble chelating agents increases the urinary excretion but lead may be concurrently mobilized from the bones. The

excretion of the PbEDTA complex is attended with some potential
danger because of the nephrotoxicity of this compound. If an individual with acute or chronic lead poisoning is given $CaEDTA^{2-}$ too
rapidly, the amount of PbEDTA transported to the kidneys may be
sufficient to cause partial or total renal failure [74,75]. This
need not be fatal if the patient can be placed on a dialysis treatment until kidney function recovers [75].

Subsequent to its use in lead poisoning, $Na_2CaEDTA$ has been
investigated as a metal-mobilizing agent in a large number of other
types of metal poisoning [5,6]. These include that due to the radioactive lanthanides [76], the actinides [77], and several other elements which form especially stable complexes with N and O donors [5].
While it is generally administered intravenously, it can be given
orally, though when given orally only a part of the dose is absorbed
[79]. The oral method of administration of this compound is avoided
because of a fear that this will enhance the absorption of any toxic
metal present in the food or inhaled as a fine dust [80].

Because EDTA forms stable complexes with such a large number
of metal ions, it has been very widely studied in biological systems
of various sorts, where it has been found to affect hundreds of processes dependent on the presence of metal ions. In recent years it
has been replaced in some of its therapeutic applications by DTPA
(see below). When EDTA is used in the treatment of lead poisoning
along with BAL, the combination of these agents seems to provide for
a more thorough de-leading of humans than either alone [5,78,79].

3.2. DTPA

Diethylenetriaminepentaacetic acid (DTPA), a higher homolog of EDTA:

Diethylenetriaminepentaacetic acid (DTPA)

forms complexes with the same type of metals as EDTA but the stability
constants of these are higher for the complexes with DTPA. Therefore
it is frequently found to be *more* effective than EDTA in those situations where EDTA has value as a therapeutic chelating agent. Once
again the calcium salt is the species commonly recommended, though
studies of Catsch and his co-workers have demonstrated the superiority
of the zinc complex for many purposes [6]. The zinc complex is a compound of very slight inherent toxicity [81] and has been proven to be
about as effective in the removal of plutonium from the mammalian body
[82] as CaDTPA. The toxic side effects of CaDTPA are sufficiently
serious to have worked against the general replacement of CaEDTA by
this compound [83,84]. Its strong tendency to tie up zinc can lead
to fetal mortality and congenital malformation when given to pregnant
rats [85], and it can impair the synthesis of DNA, RNA, and proteins
when it removes the zinc and manganese required [86]. The half-life
of injected DTPA in the human is about 3 hr [87]. Because accidents
with transuranium elements often involve dusts, CaDTPA has been also
administered as an aerosol. The LD_{50} of $Na_3ZnDTPA$ administered intraperitoneally is 17.4 mmol/kg in mice [88].

DTPA was at one time the most effective known chelating agent
for the removal of internally deposited transuranium elements in
humans [89]. In most accidents in which humans have been exposed to
transuranium elements industrially, highly specialized medical care
has been available in which the possible problems due to the side
effects of this compound have been quite well understood. There are,
accordingly, cases on record in which individuals have been treated
successfully with this compound in which the total amounts administered have been very large [93]. The zinc complex, $Na_3ZnDTPA$, is
preferred for this use at present in *clinical* cases because of its
lower toxicity [90]. DTPA equilibrates with the zinc bound to serum
protein much more rapidly than EDTA does, presumably because it possesses sufficient donor groups to tie up two Zn^{2+} ions when necessary
[91]. A very significant advantage of ZnDTPA is that for a given
level of administration of the CaDTPA, replacement with ZnDTPA allows

the number of injections to be increased approximately 30-fold before
the same toxic effect is encountered [83]. This is important because
the course of injections needed for the removal of radioactive elements from accident victims can be a long one. Since CaDTPA and
ZnDTPA are almost equally effective in the removal of the radioactive
lanthanides [89], Y and [144]Ce [92], the zinc complex is replacing the
calcium complex for such applications.

3.3. Other Aminopolycarboxylic Acids

Of the many other compounds of this type which have been prepared,
the following are examples which have been studied as possible
therapeutic chelating agents:

Triethylenetetraminehexaacetic acid (TTHA)

EGTA

cyclohexanediamine
tetraacetic acid, CDTA

Ethylenebis-N,N'-(2-o-hydroxyphenyl)glycine
(EDDHA)

All of the compounds of this class when administered at high
levels appear to be nephrotoxic, which may be related to their ability
to remove Zn^{2+}, Mn^{2+}, etc., from the kidney [94]. Of these, the ones
which have very high stability constants do show superior properties
as antidotes. Thus TTHA above is an effective antidote for acute
Cd^{2+} intoxication [95], as well as nickel [96], and other species
[97]. Because factors other than the stability constant are important
in determining the overall efficacy of an antidote, TTHA is not, for
example, a better antidote than DTPA. For the radionuclide ^{144}Ce,
TTHA was more effective in removing the cerium from the bones while
DTPA is more effective in removing it from the liver [97].

Because the $-CH_2COOH$ groups can be bonded to amines in a variety
of structures, many specialized chelating agents of this sort have
been made, e.g., those incorporated into polymers, which can function
as chelating agents. The variety of such compounds presently available can be appreciated from the number described in earlier surveys
[98,99] as well as recent, more specialized ones [100]. Because of
the considerable effort necessary to collect biological data on any
given compound of this type, only those whose properties show some
significant differences from the general run of such compounds have
been studied in animals in any detail.

4. D-PENICILLAMINE AND OTHER CYSTEINE DERIVATIVES

In D-penicillamine and other cysteine derivatives, the basic structural feature is the combination of three different types of donor
groups: -SH, $-NH_2$, and -COOH in the arrangement:

$$-\overset{|}{\underset{SH}{C}}-\overset{|}{\underset{NH_2}{C}}-C\overset{\displaystyle O}{\underset{OH}{\big\langle}}$$

The coordinating ability of this arrangement is fairly general and
such chelating agents form stable complexes with many toxic metal

ions which prefer sulfhydryl donors as well as metal ions which
prefer N and O donors. Because of the biological importance of
this basic structural type, the mammalian body possesses enzymatic
systems capable of transforming it quite rapidly. The compounds
of this type which are of most practical use have structures modified in such a manner that the rate at which they are metabolized
is reduced. A compound such as D-cysteine can furnish temporary
protection against the toxic effects of a heavy metal. In the
course of time, however, its concentration is reduced rapidly and
its antidotal action is destroyed. The structural change which has
been exploited most thoroughly is the one in which two methyl groups
are put on the carbon which bears the sulfhydryl group to give
penicillamine:

$$H_3C - \overset{\overset{\displaystyle CH_3}{|}}{\underset{\underset{\displaystyle SH}{|}}{C}} - \overset{\overset{\displaystyle H}{|}}{\underset{\underset{\displaystyle NH_2}{|}}{C}} - C \overset{\displaystyle \nearrow O}{\underset{\displaystyle \searrow OH}{}}$$

Penicillamine

4.1. D-Penicillamine

D-penicillamine was introduced by Walshe in 1956 for the purpose of
enhancing the excretion of copper from individuals suffering from a
hereditary disorder (Wilson's disease or hepatolenticular degeneration) in which they accumulated copper steadily until it caused their
death [101]. Subsequently, it has been found to be an extremely useful therapeutic chelating agent and has found application in cases
of intoxication by lead [102], mercury [103], gold [104], platinum
compounds [105], and antimony [106], as well as extensive use in
the treatment of arthritis [107]. It is a chelating agent which is
administered *orally* and, except for individuals who have an allergic
reaction, is a compound of very modest inherent toxicity. It forms
complexes of fair stability with a wide range of metals [108], though
with some, e.g., Cu^{2+}, there is a reduction which is the result of

the action of the sulfhydryl groups [109]. When administered to
individuals over a period of years, as it is in the treatment of
rheumatoid arthritis, it can lead to disorders of the immune system
[113]. Because D-penicillamine is used so widely for rheumatoid
arthritis, it is readily available.

Although information is available on the stability constants
of D-penicillamine with many metal ions, relatively few of the solid
complexes of therapeutic interest have been characterized. The LD_{50}
of D-penicillamine administered intravenously in mice is of the order
of 4000 mg/kg [110]; orally in the same species the LD_{50} value is
about 8500 mg/kg [110]. The L form of this compound is a vitamin B_6
antagonist and is correspondingly much more toxic, having an i.p.
LD_{50} of 350 mg/kg in rats [111].

When given orally, D-penicillamine enhances the excretion of
copper, but unless it is given in very large excess for a relatively
long period of time, it has relatively little effect on the net body
stores of other essential trace elements such as zinc or cobalt [112].

From a practical point of view, it seems reasonable to expect
that D-penicillamine will replace BAL in many of its applications in
the treatment of heavy metal poisoning.

4.2. Other Cysteine Derivatives

Other compounds closely related to D-penicillamine which have been
studied include N-acetyl-D,L-penicillamine, N-acetyl cysteine, homo-
cysteine and some of its derivatives, and compounds in which the same
types of donor groups are arranged differently, as in N-(2-mercapto-
propionyl)glycine.

N-acetyl-D,L-penicillamine was introduced as an antidote for
mercury poisoning in 1959 [114] for which it is quite effective.
The acetylation of the penicillamine molecule destroys most of the
donor properties of the nitrogen atom, but allows the resultant compound to penetrate cell membranes more readily. As a consequence,
N-acetyl-D,L-penicillamine can remove mercury ion and methylmercury

from tissues such as the brain much more effectively than the parent compound [115]. For this compound the LD_{50} was reported as above 1000 mg/kg [116]. It has been used in the treatment of chronic mercury intoxication in humans [117].

N-acetyl-D-cysteine has been used for the same type of situation as D-penicillamine but appears to be less satisfactory, presumably because of the greater ease with which it is metabolized in the mammalian body. It has been used to clear gold (given as gold sodium thiomalate) from tissues and is capable of enhancing the urinary excretion of gold [118] and mercury [115].

β-Methyl-β-ethylcysteine is also resistant to the action of enzymes which metabolize cysteine and has been found to increase the excretion of copper in Wilson's disease and to protect against the lethal effects of mercuric chloride in the rat [116].

N-(2-mercaptopropionyl)glycine is capable by virtue of its sulfhydryl grouping of reacting with many toxic heavy metals, such as mercury(II), and has been studied to some extent as a heavy metal antidote. Its actions seem to be rather similar to that of D-penicillamine in some ways, though it is usually not as effective as that compound in animal tests [119].

5. DESFERRIOXAMINE AND OTHER CHELATING AGENTS USED TO CONTROL IRON OVERLOAD

There are certain hereditary anemias, e.g., Cooley's anemia, which are customarily treated by frequent blood transfusions extending over a period of years. Because of the tenacity with which the human body retains its iron, this leads to a continual increase in the iron content of the liver, heart, and other organs and, eventually, to the death of the individual from iron overload. During the last 40 years, numerous studies have been carried out on the influence of chelating agents on the excretion of iron, but the problem of iron overload was not solved technically until the development of chelating agents which were rather specific for iron(III) in biological milieu.

This was achieved by the isolation of a group of compounds synthesized by microbes to extract iron from their environment. Of these the most important compound in clinical usage is desferrioxamine. This chelating agent uses hydroxamic acid groups to form very stable chelate structures with iron(III). By the intravenous administration of such a chelating agent, the urinary and fecal excretion of iron can be greatly enhanced. The fact that this compound is not given orally because of its extensive destruction in the gastrointestinal tract introduces some difficulties in the clinical control of this disorder because most victims are quite young. This has led to a clear realization of the great advantages that would result from a reasonably specific chelating agent for iron which could be administered orally.

5.1. Desferrioxamine

Desferrioxamines are naturally occurring, quite selective iron(III) chelators isolated from actinomycetes [121]. The structure of the one used to enhance iron excretion in humans is:

$$H_2N-(CH_2)_5 \overset{HO}{\underset{|}{N}}-\overset{O}{\underset{||}{C}}-(CH_2)_2-\overset{O}{\underset{||}{C}}-NH(CH_2)_5 \overset{HO}{\underset{|}{N}}-\overset{O}{\underset{||}{C}}-(CH_2)_2\overset{O}{\underset{||}{C}}NH(CH_2)_5 \overset{HO}{\underset{|}{N}}-\overset{O}{\underset{||}{C}}-CH_3$$

Desferrioxamine B

("desferrioxamine")

The parent compound is not very soluble in water and so the salt with methanesulfonic acid, which is quite soluble, is used clinically. For the reaction with ferric iron:

$$Fe^{3+} + HL^{2-} \rightleftharpoons FeHL^+ \qquad K = 10^{30.6}$$

When given intraperitoneally or intravenously, desferrioxamine complexes iron from body stores and the iron(III) complex (ferrioxamine) is readily excreted in the urine, turning it a reddish color. Desferrioxamine is not efficiently absorbed from the gastrointestinal tract so it is given intravenously; it is also subject to a combination of metabolic destruction and rather rapid excretion. It is a compound of modest inherent toxicity, the LD_{50} in mice being 1240 mg/kg for intraperitoneal administration [122] and around 10,000 mg/kg for oral administration [122]. It is also an antidote for *acute* iron intoxication [123,124].

In individuals suffering from iron storage problems subsequent to transfusions, the use of ascorbic acid with desferrioxamine considerably enhances the amount of iron excreted, though this rapid mobilization of large amounts of iron can lead to cardiac dysfunction [25].

Because the coordination chemistry and biological distribution of plutonium in the human body is very similar to that of iron [125], desferrioxamine is also useful in enhancing the rate of excretion of plutonium [120,125].

5.2. Other Chelating Agents Used for Iron Overload

The fact that desferrioxamine is not very effective when given orally has led to a search for other chelating agents for iron(III) which do not suffer from this defect. The donor arrangements which have been incorporated into these chelating agents include the hydroxamic acid groups, vicinal diphenols, and a combination of a phenol and a hydrazide.

One of these which has been used clinically is 2,3-dihydroxybenzoic acid:

which enhances primarily the fecal excretion of iron, though the urinary excretion is also increased [126]. The LD_{50} for oral administration to rabbits is in excess of 3 g/kg [127] and it is rapidly

excreted after oral administration [128]. When given to humans
orally at a level of 25 mg/kg/day *after* food it does enhance the
excretion of iron in individuals with iron storage disease [129].

Compounds of a very different type, of which the most promising
is pyridoxal isonicotinoyl hydrazone:

have also been examined for use in iron overload [130,131]. This
compound increases both the biliary and urinary excretion of iron
when given either intraperitoneally or orally [131].

A series of iron-sequestering agents whose structures are
related to the microbial iron transport agent enterobactin have been
synthesized and studied by Raymond and his collaborators [132-135].
An example is the linear trimeric sulfocatecholate ligand:

These compounds are water-soluble and have a higher affinity for iron
than desferrioxamine B [135]. The catechol groups here are the species
that bond the iron; the groups are located in these compounds so that
the hydroxy pairs on different aromatic nuclei are capable of bonding
to the same iron ion. Compounds of this type seem to have a sufficiently modest toxicity to make them of interest for tests in humans
[136].

6. OTHER CHELATING AGENTS

Many more chelating agents have been proposed for therapeutic use
than have actually been evaluated. Of those developed from purely
chemical considerations, the vast majority have had to be discarded
because they are simply too toxic. In both inorganic and analytical
applications of chelating agents it is often a distinct *disadvantage*
to have present additional polar groups which give both the chelating
agent and its metal complexes an elevated degree of water solubility.
As a result, many of the commonly used chelating agents and their
metal complexes possess a low degree of water solubility and a lipo-
philicity which lengthens the time that they are held in the body
tissues. Thus, the information that they provide on metal bonding
is of interest, but is normally useful only when it can be extended to
analogous chelating agents which possess additional water-solubilizing
groups. An exception to this is the use of sodium diethyldithiocarbamate in the treatment of $Ni(CO)_4$ intoxication [145].

Of the other chelating agents, the ones which presently seem
most interesting are those of promise in the decorporation of plutonium.

6.1. Decorporation of Plutonium

Because of the extensive manipulations involving plutonium in the
nuclear weapons industry, accidents involving this element occur with
low but seemingly regular frequency. The biological transport and
storage of this element exhibits many similarities to that of iron(III)
and this is the clue that has provided the key to the most recent
developments [5,120].

It was shown that the EDTA-type compounds, of which DTPA and
TTHA were the most effective members, are capable of significantly
enhancing the excretion of ^{239}Pu [5]. Desferrioxamine can also
mobilize ^{239}Pu from most of the organs, but tends to leave a surprisingly large amount in the kidney [5]. The use of the ZnDTPA

complex to facilitate the excretion of plutonium has become widespread and is now recognized as superior to CaDTPA because of the
lower toxicity of the zinc(II) complex [90].

The many similarities of Pu(IV) to Fe(III) [120] stimulated the
search for chelating agents with donor groups similar to those of the
naturally occurring iron-binding compounds synthesized by various
microorganisms: hydroxamic groups and ortho diphenols. Of these, the
ortho diphenols seem presently to furnish the most effective examples
of Pu(IV) binding agents. The presence of sulfonate groups also, on
the aromatic rings, gives these compounds a sufficient degree of water
solubility to allow the metal complex to be excreted. The complexes
formed with the parent (nonsulfonated) chelating agents are not excreted. The two most effective compounds were found to be of the
type:

For these compounds, which were called 4,4,4-LICAMS and 4,3,4-LICAMS,
the values of M, n, and m are:

    4,4,4-LICAMS   M = Na, n = 4, m = 4
    4,3,4-LICAMS   M = K, n = 4, m = 3

The removal of an injected dose of ^{239}Pu by these chelating
agents was approximately the same as that found for an equimolar
amount of $CaNa_3DTPA$, but the advantages of these materials were
claimed to be a lower reactivity at physiological pH with essential
divalent metal ions, a moderate toxicity, and the ability of this
type of compound to remove some of the ^{239}Pu which has already been
deposited in the bone [136]. While the toxicologic properties of
these chelating agents have not been as thoroughly investigated as
those of DTPA, their development is a striking demonstration of the
synthesis of new selective chelating agents for a toxic metal on the
basis of an effective understanding of its coordination chemistry.

6.2. Crown Ethers and Other Macrocyclic Chelators

As chelating agents, the crown ethers are outstanding for the selectivity of the complexes which they form with metal ions of the correct size and charge [137]. Unfortunately, the vast majority of such compounds are electrically neutral, so that their distribution in the body and their limited water solubility cause practical difficulties when an attempt is made to use them as antidotes. Lehn and Montavon [138] have shown that the cryptate ligand:

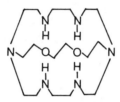

has a very high affinity for Cd^{2+}, Pb^{2+}, and Hg^{2+}. It has also been shown that it is possible to prepare a macrocycle which is very selective for cadmium(II) vis-à-vis zinc(II) or calcium(II) [139]. The effect of alterations in cavity size and the replacement of oxygen donors by nitrogen donors in such cryptate-complexing agents has been discussed in detail [139]. Of the known systems which chelate metal ions, these systems appear to be among the most readily manipulated to attain a very high degree of selectivity in their coordination to very closely related metallic ions. Many of the compounds in this group are, like ethylene glycol derivatives of all sorts, metabolized to oxalic acid. While they have been suggested for use as therapeutic chelating agents for a number of toxic heavy metals, relatively little information is available on the way they function as such in in vivo tests.

The cryptate

forms a Sr^{2+} complex which is much more stable (log K = 13) than its
Ca^{2+} complex (log K = 4.1) [140]; it has been shown to enhance both
the urinary and fecal excretion of freshly injected ^{85}Sr [141].

7. OTHER MEDICAL APPLICATIONS OF CHELATING AGENTS

7.1. Radiocontrast Agents

The injection of a very stable complex of a radioactive ion allows
the course of the species through the human body to be followed by
simply counting the disintegrations that occur and register at a
counter placed at various positions on the skin of the patient. The
larger the number of counts which occur, the easier it is to diagnose
the functioning of the organs through which the complex is moving.
Unfortunately, the use of radioactive ions with short half-lives also
make the deposition of any of that ion in the body more dangerous.
As a result, such complexes, which are often used in the evaluation
of kidney abnormalities, need to possess an optimal stability as well
as a great solubility in water. The effective stability of the complexes can be improved by injecting them in solutions that contain
an excess of the chelating agent. Subsequently any residual radioactive material can be removed by additional treatment with the
chelating agent [58].

7.2. Antimony(III) Chelates for Schistosomiasis

Arsenic(III) and antimony(III) compounds frequently are sufficiently
more toxic for lower organisms than for humans that this differential
lethality can be used in the treatment of diseases caused by these
lower organisms. The introduction of arsenic compounds for the
treatment of syphilis by Ehrlich is perhaps the best known example.
There are a number of other diseases, due to parasites, for which
this differential lethality can be used. Of these, schistosomiasis

is of interest in that one method of treatment has been via the injection of water-soluble complexes of antimony(III). Initially potassium antimonyl tartrate was used for this purpose. This compound is effective in killing the parasite but is also a compound of considerable cardiotoxicity for humans. Because some humans seem to be quite sensitive to its side effects, studies have been carried out searching for other antimony complexes which have a larger margin of safety between the therapeutic dose and the dose at which serious side effects make an appearance. The antimony(III) species coordinates to oxygen donors as in tartrate, to orthophenolic groups, as in catechol and its derivatives, and to sulfhydryl groups as in D-penicillamine, 2,3-dimercaptosuccinic acid, 2,3-dimercaptopropanol-1, etc. In general, the bonds to sulfhydryl are appreciably more stable than those to oxygen donors. A relatively large number of such complexes have been studied and it has been found that many of them possess an appreciable antischistosomal activity as well as a reduced human toxicity. They have been used in numerous clinical cases with humans and shown to be valuable in practice. They have also furnished very useful clues to the types of antidote which can be used for antimony(III) itself [142].

8. SOME CURRENT UNSOLVED PROBLEMS

There are many metal ions which, once they have been incorporated into the human body, cannot be removed or cannot be removed in a selective fashion. These may be the source of important health problems as the result of industrial or environmental contamination. In some instances, the biological half-life of these species is so long that the natural turnover rate cannot be depended upon to prevent serious consequences. One solution to such problems is the development of suitable selective chelating agents to enhance their rate of removal from the body. Two such problems of current interest are the deposition of radioactive isotopes in the bone and the storage of cadmium in the kidney.

8.1. Deposition of Radioelements in Bone

The sequence of events subsequent to the accidental ingestion of many
radioelements involves a period during which they can be removed,
with decreasing ease, by treatment with an appropriate chelating
agent; subsequently they are largely immobilized in the bone [5,6].
This immobilization in the bone is one of the methods used by the
body to detoxify metal ions such as Pb^{2+} and is a largely successful
method for ions which are not radioactive. For radioactive ions this
deposition in the bone has the unfortunate result of exposing the
bone marrow as well as other adjacent tissues to an enhanced level
of radiation. At the present time there seems to be no feasible
technique for the selective removal of such ions from their deposits
in the bone. What type of chelating agent would possess the selectivity and pharmacological properties required to is also not presently known.

8.2. Mobilization of Cadmium

When cadmium is ingested by the mammalian body it induces the synthesis of the special metal-binding protein thionein (called metallothionein when it is bound to metal ions) which bonds cadmium very
firmly in the liver, kidneys, and other organs. The cadmium is soon
bound exclusively by this material and it cannot be readily mobilized
from the metallothionein. As the cadmium content of the kidney increases, it reaches a level where kidney function is adversely
affected. This is seen most frequently in industrial workers who
handle cadmium, but is also found in individuals who live in an
environment in which the food or water has an unusually high level
of cadmium. At the present time there is no therapeutic procedure
available to physicians to remove the cadmium from such persons [143]
and there is no assurance that kidney function would return to near
normality if the cadmium were removed [143,144]. It does seem to be
the case, however, that if the cadmium is removed before it reaches
such high levels then permanent damage can be avoided.

ABBREVIATIONS

BAL	2,3-dimercaptopropanol-1
CDTA	trans-1,2-diaminocyclohexanetetraacetic acid
DFOA	desferrioxamine B
DMPS	sodium 2,3-dimercaptopropane-1-sulfonate
DMSA	2,3-dimercaptosuccinic acid
DTPA	diethylenetriaminepentaacetic acid
EDDHA	ethylenebis-N,N'-(2-o-hydroxyphenyl)glycine
EDTA	ethylenediaminetetraacetic acid
EGTA	ethyleneglycol-bis-(β-aminoethyl ether)N,N,N',N'-tetraacetic acid
LD_{50}	the amount of the compound which when given in the manner indicated to a large number of animals of the designated species will result in the death of 50% of them with a high degree of probability
TTHA	triethylenetetraminehexaacetic acid
Unithiol	sodium 2,3-dimercaptopropane-1-sulfonate

REFERENCES

1. J. Schubert, *Ann. Rev. Nucl. Sci.,* *5,* 369 (1955).

2. J. Schubert, Proc. Intl. Conf. Peaceful Uses of Atomic Energy (1st), Geneva, Vol. 13, 1955, pp. 274-297.

3. J. Schubert, *Atompraxis,* *4,* 393 (1958).

4. J. Schubert, *Fed. Proc.,* *20,* Suppl. 16, 220 (1961).

5. A. Catsch, *Dekorporierung radioaktiver und stabiler Metallionen,* K. Thiemig, Munich, 1968.

6. A. Catsch, A. E. H. Harmuth Holne, and D. P. Mellor, *The Chelation of Heavy Metals,* Pergamon Press, Oxford, 1979.

7. D. D. Perrin, *Suomen Kemistilehti,* *42,* 205-213 (1969).

8. R. P. Agarwal and D. D. Perrin, *Agents and Actions,* *516,* 667-673 (1976).

9. D. D. Perrin and R. P. Agarwal, *Metal Ions in Biological Systems,* Vol. 2, Marcel Dekker, New York, 1973, pp. 168-206.

10. P. M. May and D. R. Williams, *FEBS Lett.,* *78,* 134-138 (1977).

11. G. E. Jackson, P. M. May, and D. R. Williams, *FEBS Lett., 90,* 173-177 (1978).

12. P. M. May, P. W. Linder, and D. R. Williams, *J. Chem. Soc.* (Dalton), 588-595 (1977).

13. M. Lucassen and B. Sarkar, *J. Toxicol. Environ. Health, 5,* 897-905 (1979).

14. J. Schubert, in *Radiobiology of Plutonium* (W. S. S. Jee and B. J. Stover, eds.), University of Utah Printing Service, Salt Lake City, Utah, 1972, pp. 355-375.

15. J. Schubert and S. K. Derr, *Nature, 275,* 311-313 (1978).

16. M. M. Jones and M. A. Basinger, *Res. Commun. Chem. Pathol. Pharmacol., 24,* 525-531 (1979).

17. L. R. Cantileva, Jr., and C. D. Klaasen, *Toxicol. Appl. Pharmacol., 53,* 510-514 (1980).

18. F. Planas-Bohne, *Experientia, 36,* 1001-1002 (1980).

19. R. A. Bulman, F. E. H. Crawley, and D. A. Geden, *Nature, 281,* 406 (1979).

20. V. Volf, *Health Phys., 39,* 364-366 (1980).

21. C. W. Jones, R. D. Lloyd, and C. W. Mays, *Rad. Res., 84,* 149-151 (1980).

22. J. Schubert, *Nature, 281,* 406 (1979).

23. P. M. May and D. R. Williams, *Nature, 278,* 581 (1979).

24. V. Volf, *Health Phys., 29,* 61-68 (1975).

25. A. Nienhuis, *N. Engl. J. Med., 296,* 114 (1977).

26. M. G. Cherian, *Nature, 287,* 871 (1980).

27. L. Magos and Ts. Stoychev, *Br. J. Pharmacol., 35,* 121 (1969).

28. R. T. O'Brien, *Ann. N.Y. Acad. Sci., 232,* 221-225 (1974).

29. M. Cikrt, *Arch. Toxicol., 39,* 219-223 (1978).

30. R. A. Goyer and M. G. Cherian, *Life Sci., 24,* 433-438 (1979).

31. R. A. Peters, L. A. Stocken, and R. H. S. Thompson, *Nature, 156,* 616 (1945).

32. R. A. Peters, *Rec. Chem. Prog., 28,* 197 (1967).

33. L. A. Stocken, R. H. S. Thompson, and V. P. Whittaker, *Biochem. J., 41,* 47 (1947).

34. R. H. S. Thompson and V. P. Whittaker, *Biochem. J., 41,* 342 (1947).

35. E. S. G. Barron and G. Kalnitsky, *Biochem. J., 41,* 346 (1947).

36. A. J. Canty, in *Organometals and Organometalloids* (F. E. Brinckman and J. M. Bellama, eds.), ACS Symposium Series 82, American Chemical Society, Washington, D.C., 1978, pp. 327-338.

37. M. Berlin and T. Lewander, *Acta Pharmacol.*, *22*, 1-7 (1964).

38. M. Berlin and S. Ullberg, *Nature*, *197*, 84-85 (1963).

39. M. Berlin, L.-G. Jerksell, and G. Nordberg, *Acta Pharmacol. Toxicol.*, *23*, 312-320 (1965).

40. M. Berlin and R. Rylander, *J. Pharmacol. Exp. Ther.*, *146*, 236-240 (1964).

41. N. S. Johary and L. N. Owen, *J. Chem. Soc.*, *19*, 1307 (1955).

42. V. E. Petrunkin, *Ukr. Khim. Zhur.*, *22*, 603 (1956).

43. E. V. Vasil'eva and T. K. Nedopekin, in *Tiolovye Soedinnenia v Med, Trudy Nauchnoi Konferentsii*, 1957 (publ. 1959) (N. I. Luganskii et al., eds.), Gosmedizdat. Ukr. SSR, Kiev, pp. 36-39.

44. A. T. Pilipenko and O. P. Ryabushko, *Ukr. Khim. Zhur.*, *32*, 622-626 (1966); *Chem. Abstr.*, *65*, 11412.

45. B. Gabard, *Arch. Toxicol.*, *35*, 15-24 (1976).

46. L. N. Owen and M. U. S. Sultanbawa, *J. Chem. Soc.*, 3109 (1949).

47. Rheinpreussen Akt.-Ges für Bergbau und Chemie (H. Emde), Germ. Patent 949,055 (Sept. 15, 1956), *Chem. Abstr.*, *51*, 4418g.

48. I. M. Klotz, G. H. Czerlinski, and H. A. Fiess, *J. Am. Chem. Soc.*, *80*, 2920-2923 (1958).

49. R. Hoffman-LaRoche & Co., A. G., Belg. Patent 665,535 (Dec. 16, 1965), *Chem. Abstr.*, *65*, 3678g.

50. H. R. Stohler and J. R. Frey, *Ann. Trop. Med. Parasitol.*, *58*(4), 431-438 (1964).

51. I. E. Okonishnikova and V. L. Nireburg, *Vopr. Eksp. Klin. Ter. Profil. Prom. Intokskiatsii*, *1974*, 11-14 Ch. Abst. 84, 908m.

52. E. A. H. Friedman, J. R. DaSilva, and A. V. Martins, *Am. J. Trop. Med. Hyg.*, *3*, 714 (1954)

53. M. T. Khayyal, *Bull. WHO*, *40*, 959 (1969).

54. S.-C. Wang, K.-S. Ting, and C.-C. Wu, *Chinese Med. J.*, *84*(7), 437-439 (1965), *Chem. Abstr.*, *63*, 18928C.

55. E. Ogawa, *Igaku to Seibutsugaku*, *91*(4), 295-299 (1975), *Chem. Abstr.*, *84*, 145614X.

56. J. Aaseth and E. A. H. Freidheim, *Acta Pharmacol. Toxicol.*, *42*, 248-252 (1978).

57. C. H. Tadlock and H. V. Aposhian, *Biochem. Biophys. Res. Commun.*, *94*, 501 (1980).

58. H. D. Burns, P. Worley, H. N. Wagner, and L. G. Marzilli, *Labelled Comp. Radiopharmacol.*, *13*(2), 156 (1977).

59. J. F. Danielli, M. Danielli, J. B. Fraser, P. M. Mitchell, L. N. Owen, and G. Shaw, *Biochem. J.*, *41*, 325-333 (1947).

60. M. Weatherall, *Br. J. Pharmacol.*, *3*, 137 (1948).

61. A. Gilman, F. S. Phillips, R. P. Allen, and E. S. Doelle, *J. Pharmacol. Exp. Ther.*, *87*, Suppl. 85 (1946).

62. K. Morita, E. Noguchi, and T. Yonaga, *Japan. J. Pharmacol.*, *25*, Suppl. 58P (1975).

63. K. Morita and T. Yonaga, *Japan. J. Pharmacol.*, *25*, Suppl. 85P (1976).

64. T. Yonaga and K. Morita, *Toxicol. Appl. Pharmacol.*, *57*, 197 (1981).

65. V. E. Petrunkin, *Ukr. Chem. Zhur.*, *22*, 603 (1956).

66. V. E. Petrunkin and N. M. Lysenko, *Zhur. Obschei. Khim.*, No. 1, 309 (1959).

67. E. S. G. Barron and G. Kalnitsky, *Biochem. J.*, *41*, 348 (1947).

68. R. Pribil, *Analytical Applications of EDTA and Related Compounds*, Pergamon Press, Oxford, 1972.

69. E. Rothlin, M. Taeschler, and A. Cerletti, *Schweiz. Med. Wochschr.*, *84*, 1286 (1954).

70. M. Rubin, S. Gignac, S. P. Bessman, and E. L. Belknap, *Science*, *117*, 659 (1953).

71. E. L. Belknap, *Ind. Med. Surg.*, *21*, 305 (1952).

72. S. P. Bessman, H. Reed, and M. Rubin, *Med. Ann. D.C.*, *21*, 312 (1952).

73. K. Tsuchiya, in *Handbook on the Toxicology of Metals* (L. Friberg, G. F. Nordberg, and V. B. Vouk, eds.), Elsevier/North Holland, Amsterdam, 1979, p. 451.

74. M. D. Reuber and J. E. Bradley, *J. Am. Med. Assoc.*, *174*, 125 (1960).

75. J. Mirouze, C. Mion, P. Mathieu Daude, L. Monnier, and J. L. Selam, *Nouv. Press Med.*, *4*, 1642 (1975).

76. A. Catsch and D. K. Lê, *Strahlentherap.*, *104*, 494 (1957).

77. J. Ballou, *Health Phys.*, *8*, 731 (1962).

78. J. J. Chisholm, *J. Pediatr.*, *73*, 1 (1968).

79. H. Foreman and T. T. Trujillo, *J. Lab. Clin. Med.*, *43*, 566 (1954).

80. S. Jogo, T. Maljkovic and K. Kostial, *Toxicol. Appl. Pharmacol.*, *34*, 259 (1975).

81. F. Planas-Bohne and H. Ebel, *Health Phys.*, *29*, 103 (1975).

82. A. Seidel, V. Volf, and A. Catsch, *Int. J. Radiat. Biol.*, *19*(4), 399-400 (1971).

83. A. Catsch and E. von Wedelstaedt, *Experientia*, *21*, 210 (1965).

84. G. N. Taylor, J. L. Williams, L. Roberts, D. R. Atherton, and L. Shabestari, *Health Phys.*, *27*, 285 (1974).

85. H. Swinnerton and L. S. Hurley, *Science*, *173*, 62 (1971).

86. B. Gabard, *Biochem. Pharmacol.*, *23*, 901 (1974).

87. E. Stevens, B. Rosoff, M. Weiner, and H. Spencer, *Proc. Soc. Exp. Biol. Med.*, *111*, 235 (1962).

88. L. Washburn, private communication.

89. A. Catsch, in *Diagnosis and Treatment of Incorporated Radionuclides*, International Atomic Energy Agency, Vienna, 1976, p. 295.

90. C. C. Lushbaugh and L. C. Washburn, *Health Phys.*, *36*, 472 (1979).

91. W. Jammers and A. Catsch, *Naturwissenschaften*, *22*, 588 (1968).

92. A. Catsch, D. Kh. Lê, and D. Chambault, *Int. J. Radiat. Biol. 8*, 35 (1965).

93. N. L. Spoor, *The Use of EDTA and DTPA for Accelerating the Removal of Deposited Transuranic Elements from Humans*, NRPB-R59, National Radiological Protection Board, Harwell, England, 1977.

94. A. Catsch, *Arch. Exp. Path. Pharmak.*, *246*, 316 (1964).

95. V. Eybl and J. Sykora, *Acta Biol. Med. German.*, *16*, 61 (1966).

96. M. M. Jones, M. A. Basinger, and A. D. Weaver, *J. Inorg. Nucl. Chem.*, *43*, 1705 (1981).

97. A. Catsch and D. Schinderwolf-Jordan, *Nature*, *191*, 715 (1961).

98. S. Chaberek and A. E. Martell, *Organic Sequestering Agents*, John Wiley and Sons, New York, 1959.

99. N. M. Dyatlova, V. Ya. Temkina, and I. D. Kolpakova, *Komplexony*, Izd. Khimia, Moscow, 1970.

100. C. G. Pitt and A. E. Martell, in *Inorganic Chemistry in Biology and Medicine* (A. E. Martell, ed.), American Chemical Society, Washington, D.C., *ACS Symposium Series*, *140*, 279-312 (1980).

101. J. M. Walshe, *Am. Med. J.*, *21*, 487 (1956).

102. A. D. Beattie, *Postgrad. Med. J.* (Aug. Suppl.), 17 (1974).

103. H. V. Aposhian, *Science*, *128*, 93 (1958).

104. A. Lorber, W. A. Baumgartner, R. A. Bovy, C. C. Chang, and R. Hollcraft, *J. Clin. Pharmacol.*, *13*, 332 (1973).

105. R. Osieka, U. Brutsch, W. M. Gallmeier, S. Seeber, and C. G. Schmidt, *Deutsche Med. Wochschr.*, *101*(6), 191 (1976).

106. M. E. Tarrant, S. Wedley, and T. J. Woodage, *Ann. Trop. Med. Parasitol.*, *65*, 233 (1971).

107. W. H. Lyle, *Distamine*, Dista Prods. Ltd., Liverpool, 1974.

108. D. A. Doornbos and J. S. Faber, *Pharm. Weekblad, 103,* 1213 (1968).

109. S. H. Laurie and D. M. Prime, *J. Inorg. Biochem., 11,* 229 (1979).

110. J. B. Tu, R. Q. Blackwell, and F.-F. Lee, *J. Am. Med. Assoc., 185,* 83 (1963).

111. F. Planas-Bohne, *Z. Naturforsch., 28,* 774 (1973).

112. C. Weiner, W. E. Kollmer, and P. Schramel, *Arzneim.-Forsch., 29*(2), M. 8, 1113 (1979).

113. J. C. Crawhall, D. Lecavalier, and P. Ryan, *Biopharmaceut. Drug Dispos., 1,* 73-95 (1979).

114. H. V. Aposhian and M. M. Aposhian, *J. Pharmacol. Exp. Ther., 126,* 131 (1959).

115. J. Aaseth, *Acta Pharmacol. Toxicol., 39,* 289 (1976).

116. H. V. Aposhian, *Ann. N.Y. Acad. Sci., 179,* 481 (1971).

117. S. Z. Hirschman, M. Feingold, and G. Boylen, *N. Engl. J. Med., 269,* 889 (1963).

118. A. Lorber, W. A. Baumgartner, R. A. Bovy, C. C. Chang, and R. Hollcraft, *J. Clin. Pharmacol., 13,* 332 (1973).

119. E. Ogawa, S. Suzuki, N. Tsuzuki, M. Tobe, K. Kobayashi, and M. Hoja, in *Proceedings of the International Symposium on Thiola,* Santen Pharm. Co., Ltd., Osaka, Japan, 1970, p. 238.

120. P. W. Durbin, *Health Phys., 29,* 495 (1975).

121. J. B. Neilands, in *Inorganic Biochemistry* (G. Eichhorn, ed.), Elsevier/North Holland, Amsterdam, 1973, pp. 186-200.

122. G. Pfister, A. Catsch, and V. Nigrovic, *Arzneim.-Forsch., 17,* 748 (1967).

123. S. Moeschlin and U. Schnider, *N. Engl. J. Med., 269,* 57 (1963).

124. J. L. Robotham and P. S. Lietman, *Am. J. Dis. Child., 134,* 875 (1980).

125. D. M. Taylor, *Health Phys., 13,* 135 (1967).

126. J. H. Graziano and A. Cerami, *Sem. Hematol., 14,* 127 (1977).

127. R. W. Grady, J. H. Graziano, H. A. Akers, and A. Cerami, *J. Pharmacol. Exp. Ther., 196,* 478 (1976).

128. R. L. Jones, C. M. Peterson, J. R. Graziano, R. W. Grady, H. V. Vlassara, V. C. Canale, D. R. Miller, and A. Cerami, *Blood, 46,* 1027 (1975).

129. C. M. Peterson, J. H. Graziano, R. W. Grady, R. L. Jones, H. V. Vlassara, V. C. Canale, D. R. Miller, and A. Cerami, *Br. J. Haematol., 33,* 477 (1976).

130. P. Ponka, J. Borova, J. Neuwirt, and O. Fuchs, *FEBS Lett.*, *97*, 317 (1979).

131. M. Cikrt, P. Ponka, E. Necas, and J. Neuwirt, *Br. J. Haematol.*, *45*, 275 (1980).

132. W. R. Harris, F. L. Weitl, and K. N. Raymond, *J. Chem. Soc. Chem. Commun.*, 177-178 (1979).

133. F. L. Weitl and K. N. Raymond, *J. Am. Chem. Soc.*, *101*, 599, 2728 (1979).

134. C. J. Carrano and K. N. Raymond, *J. Am. Chem. Soc.*, *101*, 5401 (1979).

135. F. L. Weitl, W. R. Harris, and K. N. Raymond, *J. Med. Chem.*, *22*, 1281 (1979).

136. P. W. Durbin, E. S. Jones, K. N. Raymond, and F. L. Weitl, *Radiat. Res.*, *81*, 170-187 (1980).

137. A. I. Popov and J.-M. Lehn, in *Coordination Chemistry of Macrocyclic Ethers* (G. Melson, ed.), Plenum Press, New York, 1979, pp. 537-602.

138. J.-M. Lehn and F. Montavon, *Helv. Chim. Acta*, *59*, 1566 (1976).

139. J.-M. Lehn and F. Montavon, *Helv. Chim. Acta*, *61*(1), 67 (1978).

140. B. Dietrich, J. M. Lehn, and J. P. Sauvage, *Tetrahedr. Lett.*, *34*, 2885, 2892 (1969).

141. W. H. Müller, *Naturwissenschaften*, *57*, 248 (1970).

142. M. A. Basinger and M. M. Jones, *Res. Commun. Chem. Pathol. Pharmacol.*, *32*, 355 (1981).

143. K. Tsuchiya (ed.), *Cadmium Studies in Japan: A Review*, Kodansha Ltd., Tokyo, 1978, pp. 301-306.

144. J. H. Mennear (ed.), *Cadmium Toxicity*, Marcel Dekker, New York, 1979, p. 128.

145. F. W. Sunderman, Sr., *Ann. Clin. Lab. Sci.*, *11*, 1 (1981).

Chapter 4

COMPUTER-DIRECTED CHELATE THERAPY
OF RENAL STONE DISEASE

Martin Rubin
Department of Physiology and Biophysics
Georgetown University School of Medicine and Dentistry
Washington, D.C.

and

Arthur E. Martell
Department of Chemistry
Texas A&M University
College Station, Texas

1. BACKGROUND

1.1. The Problem and a Possible Solution

It has been suggested that the optimal treatment for alkaline earth
renal stone disease in humans and animals would be the oral administration of a therapeutic agent which could dissolve an existing stone
or prevent its recurrence in stone-forming patients. That this goal
has not yet been attained despite its medical importance is an im-
plicit recognition of the inherent difficulty of the problem. To
function successfully the compound would be required to transit the
polyvalent cation-laden intestine, plasma, and tissue compartments
and nonetheless arrive in the urine sufficiently free of calcium and
magnesium to achieve the desired objectives.

We have reported that the oral administration of nitrilotri-
acetate (NTA) in a quantity greater than that of the endogenous and
exogenous polyvalent cation content of the intestinal tract can dissolve existing struvite $(MgNH_4PO_4 \cdot 6H_2O)$ and brushite $(CaHPO_4 \cdot 2H_2O)$
calculi in the bladder of rats and a dog [1]. Oral NTA also inhibits
the formation of these calculi in rats. Our studies have been based
on several assumptions. Most important is an expectation that the
protein/calcium proteinate buffer of plasma which maintains the
critically important calcium ion concentration could be replaced by
a suitable chelate. Glomerular filtration of the small-molecular-
weight chelate and calcium ion would present the reabsorptive surface
of the nephron with a potentially mobile equilibrium. The normal
tubular reabsorption of essentially all the filtered calcium ion
could then be expected to result in the dissociation of chelate
buffer with release and reabsorption of additional calcium ion.
Depending on the extent of this overall process a relative excess
of calcium-free ligand would be present in the excreted urine. Our
computer calculations of the speciation of chelates in intestine,
plasma, and urine indicate that a ligand with the calcium-binding
characteristics of NTA would be appropriate. In addition, its
other favorable physicochemical, metabolic, and pharmacological

and as the hydroxy and carbonate appatites, $[Ca_{10}(PO_4)_6(OH)_2]$ and $[Ca_{10}(PO_4)_6(CO)_3]$, respectively, are especially frequent in admixture with struvite, $MgNH_4PO_4 \cdot 6H_2O$, in alkaline urines. Sodium and potassium occur in concentrations up to 3 and 0.5%, respectively. The extensive list of elements which may be found in trace concentrations include aluminum, copper, iron, lead, tin, zinc, chlorine, and sulfur.

1.3. Urine Ion Activity, Supersaturation, and the Etiology of Stone Formation

Significant progress in attaining an understanding of the quantitative relations governing the precipitation of the stone-forming substances in urine was initiated by Meyer [10] and accelerated markedly in mid-century with the studies of Prien and co-workers [11]. By establishing the ionic concentrations at which urine supersaturation occurs at various pH levels and the conditions requisite for initiation of solid-phase separation, the stage was set for the more rigorous application of physicochemical principles to this process. Based on measured ionic concentrations, Robertson et al. [12] calculated absolute activity products in urine. In analogous studies Finlayson and Miller [13] correlated clinical observations with experimental and theoretical studies by treating urine as a multielectrolyte system of interactive charged species subject to stability constants appropriate for various ambient conditions. In this way it was possible to utilize iterative numerical computer calculations to investigate the potential effects of change on the system stability.

A number of factors influence the ionic activity and potential supersaturation of urine. Dehydration and disease can reduce the normal urine excretion of 600-2500 ml/24 hr of the adult human with consequent increase in solute concentration. The range of urine acidity is from 4.6 to 8.0 with usual values between 5.5 and 6.5. Numerous studies and computer calculations confirm that the precipitability of calcium oxalate is essentially independent of pH in the acidic physiological range. On the other hand, brushite forms

between pH 5.9 and 6.9, apatite between 6.6 and 7.8, and struvite
between 7.2 and 8.5. Normal urinary calcium excretion may vary from
100 to 300 mg/24 hr (2.5-7.5 meq) with some decrease possible when
oral calcium intake is diminished. Increases in calcium excretion
may reflect elevations in dietary calcium or can be associated with
a wide spectrum of disease. While hypercalciuria, an increased
urinary excretion of calcium, is a common finding in calcific stone
disease, it is not a consistent occurrence. The normally low level
of urine oxalate concentration can increase with a diet rich in vege-
tables, especially spinach and rhubarb. Massive increases occur in
a hereditary disorder of glycine metabolism in which this amino acid
is shunted into a pathway leading to oxalate as the end product.
Ingestion of ethylene glycol is rapidly converted to oxalate and
deposited as calcium oxalate in the renal tract. Pyridoxine deficiency and large doses of ascorbic acid also result in enhanced
oxalate excretion. The relation of these two factors to clinical
disease is uncertain. Phosphate, while nominally excreted in the
normal range of 0.7-1.6 g (20-50 mM), expressed as phosphorus/24
hours, varies significantly with diet and metabolism, decreasing
with increased carbohydrate utilization. Changes are also associated with intestinal problems resulting in decreased absorption, in
circumstances of high tissue demand such as pregnancy, and in renal
disease. The normal excretion of magnesium is from 100 to 200 mg/
24 hr (8-16 meq) and varies with nutrition. The excretion of NH_4^+,
of most significance for struvite formation, can vary normally
between 0.4 and 1.0 M (30-70 meq). Its output is related to urine
acidification which in turn is dependent on nutrition, physiology,
and disease. The formation of struvite calculi in humans has a
clear relation to the alkalinity of urine. In persistent infection
by urea-splitting organisms the consequent major increase in the
formation of ammonia results in an alkaline urine of high ammonium
ion concentration. This establishes the conditions favoring the
formation of struvite calculi.

Although urine saturation or supersaturation is a necessary
requirement for solid-phase formation, it is not the sole determinant

of whether crystal formation and crystal growth or aggregation will result in stone disease.   In the absence of urinary stasis, the frequent periodic emptying of the bladder can serve to protect against crystal formation and the slow growth of calculi to a size that cannot be spontaneously excreted. Thus despite the presence of frequent crystalluria in the urine of normal individuals and its increase in stone formers, this protective physiological mechanism guards against stone formation. While homogeneous induction of crystallization can occur in supersaturated solutions, this is not considered to be of potential importance in stone disease. More likely possibilities for the provision of a heterogeneous focus of crystallization include the normal presence of cellular debris arising from epithelial surfaces lining the renal tract, the presence of bacteria, of unique urinary proteins, and of compounds whose similarity in crystal lattice structure can provide the potential for epitactic crystal induction and growth. Octacalcium phosphate, brushite, the apatites, uric acid, and urates have been proposed for such a role. On the other hand, a number of normally occurring urine constituents serve as inhibitors of crystal formation and growth. The most significant of these, pyrophosphate, apparently functions by adsorption on the crystal-forming surface. At a normal urine concentration of 10^{-5}-10^{-4} M it has an unambiguous capability to inhibit crystal formation in a number of systems. While the concentration of pyrophosphate is decreased in some stone-forming individuals, this is not a uniform finding. Thus the demonstrable inhibitory properties of citrate, phosphocitrate, magnesium, zinc, and other trace elements may also play an individualized and significant role in the propensity toward crystal formation. An increased tendency for the aggregation of calcium oxalate and calcium phosphate crystals into larger clusters is characteristic in the urine of individuals forming such calculi. Their aggregation is inhibited by pyrophosphate, citrate, and by some glycosaminoglycans, notably heparin, hyaluronic acid, and dermatan sulfates. The etiological role of these materials in stone disease is uncertain.

1.4. Therapy of Stone Disease

Therapeutic efforts directed toward the dissolution and the prevention of recurrent calculi formation have been responsive to the
accumulating knowledge of their composition and pathogenesis. An
early and continuous recommendation has been that patients maintain
a high fluid intake to produce an undersaturated diluted urine. While
there are experimental studies and a suggestion of clinical evidence
to indicate that urine unsaturation can cause the dissolution of struvite calculi, this process has little potential applicability to other
calculi. It does appear beneficial in decreasing the incidence of
recurrent calculi. In practice, however, it has been difficult to
sustain the required copious fluid intake for the protracted time
required for treatment. Even though calcific stone formation occurs
in individuals with normal calcium urinary concentration as well as
in persons with hypercalcinuria, an obvious approach to therapy has
been the effort to reduce calcium excretion in urine. Dietary calcium
restriction to reduce intake and various means to reduce intestinal
absorption of calcium have been explored. The oral administration
of 10-15 g/day of cellulose phosphate, which binds divalent ions,
reduces urine calcium but increases oxalate excretion. Nonetheless,
it is reported to decrease the incidence of stone formation. Phytic
acid with an analogous function is partially hydrolyzed by intestinal
phosphatases releasing phosphate for absorption and urinary excretion.
The lack of palatability, large bulk, and tendency of these compounds
to cause diarrhea has limited their clinical acceptability. Calcium
absorption can also be readily decreased by oral oxalate. As in the
case of other calcium-binding compounds, the calcium complexes are
excreted in the feces. This process, especially in the case of oral
oxalate therapy, is characterized by enhanced absorption and excretion
of oxalate in the urine. It has also been pointed out that calcium is
an essential nutrient and a long-term restriction of its absorption
could have negative physiological consequences. The oral administration of thiazide drugs provides an alternative mechanism to reduce
urinary calcium. They have a direct action on renal excretion of

calcium with simultaneous increase in magnesium excretion of 20-30%
and increase in zinc excretion of 50%. In addition, the long-term
therapy with these compounds decreases urinary oxalate excretion.
This latter result appears to be a secondary one perhaps related to
a decreased need for intestinal calcium absorption, in turn causing
an increase in formation and fecal excretion of intestinal calcium
oxalate. The use of thiazide drugs is stated to have a positive
result in the treatment of calcific stone disease. The observation
that oral orthophosphate administration could reduce urinary calcium
in hypercalciuric patients and the studies demonstrating the potent
inhibitory action of pyrophosphate on crystal formation and aggregation stimulated efforts to apply these findings to the problem of
stone formation. Positive results have been reported in an extensive
series of patients treated with orthophosphates. This favorable outcome has been attributed to an increased excretion of urine pyrophosphate and other hydrolyzable inhibitors of mineralization. It also
causes an increased excretion of citrate and possibly "the citrate-
like inhibitor." Oral administration of a polyphosphate, presumably
hydrolyzed to phosphate in the intestine, was equally effective. The
organic phosphonates ethane-1-hydroxy-1,1-diphosphonate and dichloro-
methylene diphosphonate are also potent inhibitors of the formation
and aggregation of calcific calculi in vitro and in experimental
animals. Their utilization in humans has been discouraged because
the large doses required to reduce crystalluria are sufficient to
cause bone disease. While the procedures which have been summarized
may have some potential for influencing the incidence of recurrent
calcium stone disease, none appear to have the potential for dissolving existing calculi.

The alkaline environment of high ammonia concentration favorable to the formation of struvite calculi would appear to be amenable
to appropriate manipulation for the therapeutic control of struvite
formation and solution. Indeed the combination of the administration
of acetohydroxamic acid, a urease inhibitor, with appropriate antibiotics and restriction of dietary phosphate supplemented with Basogel,
a phosphate-binding compound, appears to have had a favorable influence

on the disease process. Alternatively, urinary acidification by
various means has been utilized in an attempt to reduce the urine pH
to the point where brushite calculi would dissolve and the inhibition
of their formation would occur. Neither approach has had more than
sporadic success. The presence of an existing stone in an infected
host sets up a vicious cycle in which the eradication of infection
is difficult and the facile dissolution of the stone impeded. Likewise, the continued long-term maintenance of an acidic urine in these
circumstances is fraught with difficulty. Nonetheless, the combination of surgery, when indicated, with the medical therapy described
above remains the treatment of choice for this form of stone disease.

The minimal success in achieving the dissolution of existing
calculi by drug therapy or other systemic efforts lends some interest
to alternative attempts to achieve calculus dissolution. Thus retro-
grade irrigation of the bladder and ureter with solutions capable of
achieving this objective has been utilized with modest success [14].
Suby proposed and studied the utilization of a citric acid solution
at pH 4 for this purpose [15]. While the in vitro solution of calcium
phosphate calculi could be demonstrated, it had little effect on calcium oxalate stones. It has not had acceptance as a useful therapeutic modality. Neutral solutions of ethylenediaminetetraacetate (EDTA)
are more effective both in vitro and in vivo but the intense pain and
hematuria associated with their clinical use has discouraged this
approach [16].

2.   ORAL CHELATES IN THE THERAPY
OF RENAL STONE DISEASE

2.1.   Chelates and the Intestinal Compartment

Factors which determine the oral availability of chelating ligands
include the heterogeneity of ambient conditions in the gastrointestinal tract, its luminal composition in relation to the kinetics of
intestinal passage and the formation of metal chelates, their potential metabolic modification and intrinsic absorbability. This

discussion considers the general class of aminopolycarboxylate
ligands with specific focus on EDTA and NTA as pertinent examples.
The pH of the human intestine varies sharply from the strong acidity
(pH 1-2) of the stomach to the modest acidity or mild alkalinity characteristic of the absorptive area of the upper intestine. Whether
the role of stomach acidity in serving to cleave most normally occurring metal complexes will be equally applicable to synthetic metal
chelates has not been established. Nor is there information as to
whether stomach acidity and the kinetics of the passage of stomach
contents into the upper intestinal tract permit the precipitation of
acid forms of either NTA or EDTA in the stomach. In vitro data suggest that such cleavage and precipitation in the stomach would not
be an unreasonable outcome. Were this indeed the fact, then the
metal-binding equilibria reestablished in the upper intestine would
depend on the pH, the concentration of reactants, the rapidity of
metal-ligand interaction in relation to the intestinal absorbability
of possible species, and the rate of passage of constituents across
the absorptive mucosal surface. It is known that oral EDTA is poorly
absorbed in animals and humans [17]. It serves as a calcium trap with
resulting enhanced fecal calcium excretion [18]. On the other hand,
there is evidence for the oral absorption and subsequent renal excretion of lead EDTA [19]. Oral NTA, in contrast, is essentially completely absorbed in rats and, to a lesser degree, in other species
including humans [20].

As we have pointed out, however, some studies may have over-
looked the important consequences of the relative concentrations of
polyvalent cations and other ligands on the speciation of NTA [21].
This calculation has been facilitated in our work by the use of a
general-purpose computer program adapted to the study of acid-base
coordinative interactions and dissolution and precipitation in aqueous
systems [22]. When applied to the intestinal compartment, for example,
it has permitted the calculation of oral dose of NTA required for addition to the animal feed to assure an excess of the metal-free ligand.

2.2. Chelates and the Plasma Compartment

Even though plasma is the most complex matrix known, certain of its
characteristics facilitate calculation of the pertubations induced
by the introduction of chelates into this compartment. The pH is
maintained within narrow limits close to 7.4 and the concentrations
of many constituents are regulated precisely by homeostatic mechanisms. Calcium ion at 1.25 x 10^{-3} M represents close to 50% of the
total calcium of the plasma with the balance (other than 1-2% which
occurs as citrate and other soluble complexes), combined with plasma
protein, especially albumin. The introduction of EDTA reduces the
ionic calcium acutely with the formation of equivalent quantities of
calcium EDTA [23]. This application is of medical utility in the
temporary alleviation of the effects of hypercalcemia [24]. Homeostatic mechanisms, presumably through the release of parathyroid
hormone, bring about repletion of the plasma calcium from skeletal
reserves and by decreased renal excretion. The calcium EDTA is
excreted in the urine [25]. Because the calcium binding of NTA in
plasma is in the range of the existing plasma calcium buffer system
its introduction into this compartment provides less perturbation of
ionic calcium levels. As for all buffer systems its efficiency
depends on the relative concentrations of the cation and the ligand.
A large influx of ligand more rapid than the capability for calcium
replacement will cause a hypocalcemic response. However, within
physiologically acceptable levels which trigger only a modest decre-
ment in ionic calcium levels, the NTA/calcium NTA buffer pair will
be the predominant chelate species. The ligand and variable quantities of accompanying calcium are filtered by the kidney and excreted
in the urine [26].

2.3. Chelates and the Kidney

Small-molecular-weight plasma constituents are filtered by the kidney
glomerulii and proceed through the renal nephron to urine. Some,

such as calcium, are partially or completely reabsorbed by the nephron. The reabsorption of filtered calcium is essentially 99% complete primarily in the proximal portion of the nephron, although a small but significant quantity is reabsorbed in the more distal portions. The normal urinary calcium excretion is in the range of 100-300 mg/24 hr. Comparison of the excretion of calcium after intravenous administration to humans of EDTA and its preformed calcium EDTA chelate revealed that less calcium was excreted after infusion of the calcium-free chelating agent [25]. There is an anomaly in this difference because, as appears clear, the infused EDTA is immediately converted to the calcium chelate. Thus there should be little difference in the total urinary calcium excretion. A possible explanation is provided by a sequence of events starting with increased parathyroid hormone output stimulated by EDTA hypocalcemia followed by the characteristic action of parathyroid hormone in inducing increased calcium reabsorption by the nephron. The mechanism of this parathyroid hormone function has not been completely elucidated. Consideration, however, of the possibilities that can occur during the progress of calcium EDTA along the nephron is revealing. Filtration of calcium EDTA at the glomerulus and the continuation of the maintenance of pH 7.4 in the proximal tubule does not provide an opportunity for dissociation of calcium EDTA and reabsorption of calcium in this region. This follows since the concentration of calcium ion in equilibrium with EDTA at this pH would be two orders of magnitude greater than that remaining after the normal reabsorption of calcium in this portion of the nephron. In the distal nephron, however, where renal localization of parathyroid hormone takes place, acidification of urine occurs and the final control of the excretion of calcium in urine is exerted. This combination of events points to a possible mechanism for the release of calcium from EDTA and further reabsorption of the ion since dissociation of calcium EDTA increases with increasing acidity. Some support for this possibility is provided by the observation that the administration of Diamox, a carbonic anhydrase inhibitor, eliminates the differential excretion of calcium in rats following the administration of EDTA and calcium EDTA [26]. This drug is effective in

the distal portion of the tubule inhibiting the reversible hydration
of CO_2 to carbonic acid, in turn resulting in increased sodium excretion and alkaline urine pH which would favor decreased dissociation
of calcium EDTA.

The evidence that the dissociation and reabsorption of calcium
from calcium chelates can occur during their transit along the nephron
has important implications for the potential utilization of systemic
chelation in the therapy of urolithiasis. In the case of the filtered
NTA/calcium NTA buffer pair the reabsorption of calcium ion in the
proximal tubule can be expected, because of the relatively low binding
constant for the chelate equilibrium, to enhance the dissociation of
calcium NTA and the further reabsorption of calcium ion. In addition,
any stimulation of parathyroid hormone output would facilitate this
outcome. The net effect would be an increased ratio of the excretion of NTA in the urine in relation to the total calcium. The
magnitude of such a result would depend on the kinetics of dissociation, the capability for reabsorption of calcium by the nephron, the
rate of passage of the chelate along the nephron, its concentration,
and perhaps additional factors. We have explored some of these variables by intravenous injection of labeled NTA in rats with measurement
of the rate of its excretion in urine with the associated output of
calcium. The data (Fig. 1) clearly demonstrate that there is a time-

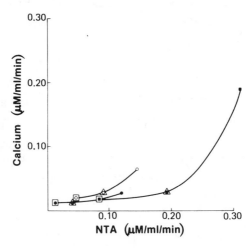

FIG. 1. Relation of urine excretion of NTA and calcium after i.v.
administration to rats: ■, ○, ●, 81.5, 40.8 and 13.6 mg/kg, first
hour, △ second hour, □ third hour, respectively.

dependent ability of the kidney to cleave calcium from excreted
calcium NTA in its passage through the kidney. The increased efficiency of this process by the end of an hour and an observed massive
phosphaturia which is the hallmark of increased parathyroid hormone
excretion points to the operation of a mechanism such as has been
proposed previously in the case of EDTA. The data also suggest that
there may be a maximum in the operative capability of the kidney for
such a system of calcium reabsorption. For application to the problem
of oral therapy of renal stone disease it is clear (Fig. 2) that the
attainment of an optimal ratio of the urinary excretion of NTA/calcium
will require careful consideration of the relation of oral intake and
renal excretion [27]. The data obtained in these acute studies confirm and extend results obtained in rat feeding. Anderson and Kanerva
[28] reported that urinary calcium levels were similar to those of control animals when the NTA ingestion was less than 600 μM per 100 grams
body weight per day. Above this threshold there was a dose-dependent
increase in urinary NTA and calcium. At feeding levels of 0.75% of
NTA crystalluria consisting of NaCaNTA was observed. In considering
the aqueous solubility of this salt, consideration needs be given to
the observation that the solubility is increased with increasing ionic

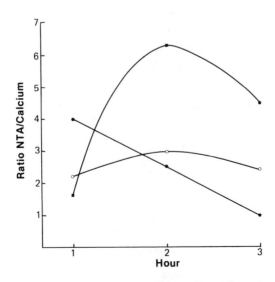

FIG. 2. Ratio of urine excretion of NTA/calcium after i.v. administration to rats: ■, ○, ●, 81.5, 40.8, 13.6 mg/kg.

strength, phosphates, pyrophosphate, calcium ion, and manganese.
Solubility decreases sharply on either side of pH 6.5 [29].

2.4. Chelates and the Urine

Our computer calculations of the effect of the introduction of NTA
into the urine compartment provided an expectation, confirmed by
experiment, of the possibilities and the limitations inherent in the
use of this ligand. At neutral to alkaline pH it was capable of dissolving mixed brushite and struvite calculi in vitro within 48 hr.
Following oral administration mixed in the feed of rats at a concentration of 0.5%, it was capable of dissolving these calculi formed
on bladder-implanted zinc sheets. It was possible to demonstrate
by x-ray and clinical evidence that its oral administration resulted
in the dissolution of a phosphatic stone in a dog. The formation of
bladder stones in rats could also be prevented. As predicted it was
ineffective in inhibiting the formation of ethylene glycol-induced
oxalate calculi [1].

The use of NTA as a substitute for polyphosphates in detergents
has resulted in detailed study of its pharmacology and toxicology.
Following its species-dependent oral absorption previously cited,
the compound is localized to the extracellular water, primarily the
plasma, is not metabolized, and is rapidly excreted in the urine.
There is no evidence of its reabsorption at the kidney [20]. A minor
fraction appears to be adsorbed to bone [30]. Significant toxicity
associated with renal vacuolization and development of malignancy
occurs with increasing frequency on lifetime feeding at levels greater
than 0.5%. The mechanism of the toxicity is unclear but its association with NaCaNTA crystalluria has been noted [31].

3. GENERAL CONCLUSIONS

The oral administration of chelating ligands provides a possibility
of a new therapeutic approach for the treatment of renal stone disease.

The computer calculation of speciation in body compartments permits the determination of chelate-introduced changes which may be within physiologically tolerable limits. In a test of this approach we have calculated the chelate stability constant required to establish a chelate-calcium buffer in plasma, considered the probable consequences of its metabolism and renal excretion, and considered its effect on the ion activity and calculi formation in urine. In keeping with prediction, NTA was demonstrated to favorably influence the formation and dissolution of mixed brushite and struvite calculi, which are the major constituents of some animal stones. Oxalate calculi, the predominant species in humans, were unaffected by the therapy, as anticipated. It can be expected that systematic exploration of the possible therapeutic role of oral chelates in the treatment of renal stone disease will be rewarding.

REFERENCES

1. M. Rubin, R. Gohil, A. E. Martell, R. J. Motekaitis, J. C. Penhos, and P. Weiss, in *Inorganic Chemistry in Biology and Medicine* (A. E. Martell, ed.), ACS Symposium Series 140, American Chemical Society, Washington, D.C., 1980, p. 381.

2. J. S. Klausner and C. A. Osborne, in *Current Veterinary Therapy, Vol. 7, Small Animal Practice* (R. W. Kirk, ed.), W. B. Saunders Company, Philadelphia, 1980, p. 1168.

3. R. W. Greene and R. C. Scott, in *Textbook of Veterinary Internal Medicine,* Vol. 2 (S. J. Ettinger, ed.), W. B. Saunders Company, Philadelphia, 1975, p. 1547.

4. D. C. Blood, J. A. Henderson, and O. M. Radostits, *Veterinary Medicine,* Lea and Febiger, Philadelphia, 1979, p. 288.

5. C. Lagergren, *J. Urol., 87,* 994 (1962).

6. W. H. Boyce, F. K. Garney, and H. E. Strawcutter, *J. Am. Med. Assoc., 161,* 1437 (1956).

7. L. C. Herring, *J. Urol., 88,* 549 (1962).

8. M. Rubin, N. J. Boyd, C. Milton, G. Davidson, W. S. Andrus, and E. J. Dwornik, in *Calcium Binding Proteins* (F. L. Siegel, E. Carafoli, R. H. Kretsinger, D. H. MacLennan, and R. H. Wasserman eds.), Elsevier/North Holland, New York, 1980, p. 501.

9. J. B. Lian, E. L. Prien, Jr., M. J. Glimcher, and P. M. Gallop, *J. Clin. Invest., 59,* 1151 (1957).

10. J. Meyer, *Ztschr. f. klin. Med.*, *111*, 613 (1929).

11. E. L. Prien, Jr., *Ann. Rev. Med.*, *26*, 173 (1975).

12. W. G. Robertson, M. Peacock, and B. E. C. Nordin, *Clin. Sci.*, *34*, 579 (1968).

13. B. Finlayson and G. H. Miller, Jr., *Invest. Urol.*, *6*, 428 (1969).

14. W. P. Mulvaney, *J. Urol.*, *82*, 546 (1959).

15. H. I. Suby, *J. Urol.*, *68*, 96 (1952).

16. B. S. Abeshouse and T. Weinberg, *J. Urol.*, *65*, 216 (1951).

17. H. Foreman and T. T. Trujillo, *J. Lab. Clin. Med.*, *43*, 566 (1954).

18. R. O. Thomas, T. A. Litovitz, and C. F. Geschickter, *Am. J. Physiol.*, *176*, 381 (1954).

19. M. Rubin and G. Di Chiro, *Ann. N. Y. Acad. Sci.*, *78*, 764 (1959).

20. W. R. Michael and J. M. Wakim, *Toxicol. Appl. Pharmacol.*, *18*, 407 (1971).

21. M. Rubin and A. E. Martell, *Biol. Trace Elem. Res.*, *2*, 1 (1980).

22. F. Morel and J. Morgan, *Environ. Sci. Technol.*, *6*, 58 (1972).

23. A. Popovici, C. F. Geschickter, A. Reinovsky, and M. Rubin, *Proc. Soc. Exp. Biol. Med.*, *74*, 415 (1950).

24. H. Spencer, J. Greenberg, E. Berger, M. Perrone, and D. Laszlo, *J. Clin. Lab. Med.*, *47*, 29 (1956).

25. H. Spencer, V. Vankinscott, I. Lewin, and D. Laszlo, *J. Clin. Invest.*, *31*, 1023 (1952).

26. M. Rubin and G. E. Lindenblad, *Ann. N. Y. Acad. Sci.*, *64*, 337 (1956).

27. D. Oday, Ph.D. thesis, Georgetown University, Washington, D.C. (1983).

28. R. L. Anderson and R. L. Kanerva, *Food Cosmet. Toxicol.*, *64*, 337 (1956).

29. K. L. Cheng and E. Lin, *Mikrochim. Acta*, *1*, 227 (1976).

30. J. A. Budny, *Tox. Appl. Pharmacol.*, *22*, 655 (1972).

31. R. L. Anderson, *Food Cosmet. Toxicol.*, *18*, 65 (1980).

Chapter 5

DETERMINATION OF TRACE METALS IN BIOLOGICAL
MATERIALS BY STABLE ISOTOPE DILUTION

Claude Veillon
U.S. Department of Agriculture
Beltsville Human Nutrition Research Center
Beltsville, Maryland

and

Robert Alvarez
Office of Standard Reference Materials
U.S. National Bureau of Standards
Washington, D.C.

1. INTRODUCTION

1.1. Isotope Dilution

The majority of the elements in the periodic system consist of two
or more stable (nonradioactive) isotopes, having different atomic
masses but identical chemical properties. In the United States,
most of the stable isotopes of the metallic elements are commercially
available in purified (enriched) form from the Oak Ridge National
Laboratory (Oak Ridge, Tennessee). With these enriched isotopes, it
is possible to determine by several methods trace metals in a variety
of materials using the technique of isotope dilution. We will concentrate herein on the application of this technique to the determination of metals at low concentrations in biological materials and
the use of stable isotopes as metabolic tracers.

A vast and important field employing stable isotopes of non-
metals also exists. Here we mean the application of stable isotopes
of elements such as hydrogen (^{2}H), carbon (^{13}C), nitrogen (^{15}N), and
oxygen (^{18}O) in the investigation of metabolism, fate, or uptake of
labeled compounds in such fields as nutrition, medicine, and agriculture. These applications have recently been summarized in an
excellent review by Bier [1].

The concept of isotope dilution for trace metal determinations
is both simple and elegant. In essence, a known quantity of an enriched stable isotope is added to the sample to be analyzed. Com-
paring, by whatever method is used, the amount of that isotope (added)
relative to another isotope (not added) of the element, one can calculate the amount of analyte present originally in the sample. In
other words, the normal relative amounts of the two isotopes (natural
abundance) has been altered by the addition of the enriched isotope;
its normal isotopic distribution has been "diluted."

1.2. Advantages

Stable isotope dilution can be an extremely powerful analytical method for trace metal determinations, with several important advantages over other methods. In the discussions which follow, it should be kept in mind that the comments apply to any method capable of measuring individual isotopes. Mass spectrometry is the most widely used method, so the discussions are primarily in terms of this measurement technique. Also, the discussions are primarily in terms of measuring stable isotopes in biological materials, which often requires sample manipulations such as destruction of the organic matter and extraction of the analyte prior to analysis. We shall assume for the moment that a known amount of an enriched isotope of the element in question (spike) has been added to the sample, that subsequent chemical processing renders the spike and endogenous analyte in the same chemical form, and that contamination is under control.

One major advantage of the isotope dilution method is that it does not depend on quantitative recovery of the analyte. If a portion of the analyte is lost through some means in sample processing, the same portion of the spike will also be lost, since they are indistinguishable chemically. Since the determination involves the relative amounts, e.g., isotope ratio, of two isotopes, the results will be unaffected. This elimination of the usual requirement of quantitative or known chemical recovery in other methods is an important advantage of the isotope dilution method.

An equally important advantage of the method is that it is absolute, i.e., calibration with samples of known concentration is not required. The spike represents an ideal internal standard because it is chemically identical to the analyte and would be affected to the same extent by any interferences. For example, in a trace metal determination by atomic absorption spectrometry, the instrumental

response must be calibrated with standards, and matrix effects caused
by other things present in the samples can often alter the results.
Unless standards and samples have identical matrices, or some technique like the method of additions is used, the results may be in
error. With isotope dilution, as long as sufficient material is
recovered to allow the measurement, the result is correct.

The accuracy of the isotope dilution method thus hinges almost
entirely on the accuracy with which the amount of added spike is
known. By way of example, let us say that our spike is a solution,
aliquots of which are accurately pipetted into samples. The concentration of this solution can be readily determined with a high degree
of certainty by inverse isotope dilution, i.e., using one or more
accurately prepared solutions of the natural abundance element and
treating the spike solution as an unknown. If the measurement is
made on the same mass spectrometer to be used for later measurements,
any mass discrimination by the instrument is cancelled out.

Thus, the primary advantages of isotope dilution methodology
over other trace metal analysis methods are the relative freedom from
the constraints of quantitative analyte recovery and the high degree
of accuracy readily achievable. Some other advantages include high
specificity, wide dynamic range, excellent sensitivity, and usually
very good precision with most measurement methods. Approximately
70 elements could, in theory, be determined by isotope dilution.
These elements have two or more stable isotopes; a few have long-
lived radioisotopes which could also be used in a similar way.

While on the subject of advantages, and concerning trace metals
in biological materials, we wish to point out a perhaps obvious advantage regarding an application of stable isotopes. In addition to
using isotope dilution to determine total analyte concentration in
samples, stable isotopes can also be used as tracers, just as radioisotopes can. For example, if one wishes to follow the metabolic
fate of a particular trace element in laboratory animals, one could
give a radioisotope of that element which could then subsequently be
easily measured in samples. However, for a similar experiment in

humans, the use of a radiotracer might be contraindicated due to the
radiation hazards. This would be particularly true in the case of
children, pregnant women, or women of child-bearing age. A stable
isotope could be used in this case, eliminating the radiation hazards.
It would also have the advantage of an "infinite" half-life, permitting long-term studies. Obviously, this would assume the availability
of a stable isotope of that element. A stable isotope would also be
used in situations where a suitable radioisotope did not exist. This
advantage will be discussed in more detail in a later section.

1.3. Disadvantages

Perhaps the biggest limitation encountered in isotope dilution is the
fact that about 17 elements are monoisotopic, i.e., only one isotope
of the element is present in nature. Some of these are important or
of interest biologically, such as F, Na, Al, P, Mn, Co, and As. For
these, alternate analytical methods must be used.

For tracer studies and other applications in which radioisotopes
can be used, using stable isotopes is far less convenient and more
time consuming. Even for routine trace metal determinations of single
or multiple elements in biological materials, some other analytical
methods are usually faster, more convenient, and require simpler
instrumentation. Aside from tracer studies, the isotope dilution
methods perhaps serve best to verify the accuracy of these other
methods.

Contamination of the samples with analyte impurities is, of
course, a problem equally common to all trace element analytical
methods. The problems of controlling and eliminating contamination
are basically the same for all methods, differences lying in the
specific procedures, reagents, etc., required in each method. One
type of contamination more important to isotope dilution methods is
to be avoided at all cost. This would be to accidently contaminate
the laboratory and/or apparatus used with an *enriched* isotope of the
analyte, which could subsequently contaminate samples. Contamination

with normal abundance material could be systematically identified
and corrected, whereas contamination with material of altered isotope
ratio would be a considerably more serious problem.

2. METHODS USED

2.1. Spectroscopic Methods

Spectroscopic methods capable of measuring stable isotopes include
atomic emission, absorption and fluorescence spectrometry, and
Mössbauer spectrometry. The first three are based on resonance
radiation, usually transitions of an electron in the atom to and
from the ground electronic energy level. Under high resolution, the
spectral lines of numerous elements are split into hyperfine components, due to differing nuclear mass and to increasing nuclear size.
The mass effect results in decreasing hyperfine component separation
with increasing atomic mass, while nuclear volume effect results in
increasing separation with increasing nuclear size, so the two effects
act in opposite directions. Consequently, one finds relatively large
isotopic line separation only for very light and very heavy elements,
with midrange elements having very small separation. Although not by
isotope dilution nor in biological materials, isotopic determinations
by these three spectroscopic techniques have been reported for heavy
elements such as U [2], Pb [3], and Hg [4,5], and for light elements
such as Li [6,7] and B [8]. Veillon and Park [5] also looked at In,
Zn, and Cu by atomic fluorescence and found the hyperfine splitting
insufficient.

The principal advantage of these spectroscopic techniques lies
in the relative simplicity of the instrumentation required. This is
perhaps offset by the fact that only isotopes of the very light and
very heavy elements can be observed, and these are not usually of
great biological interest.

Another spectroscopic technique which can be utilized to
observe stable isotopes of a few elements with certain nuclear properties is Mössbauer spectroscopy. In the case of iron, employing

^{57}Fe, Mössbauer spectroscopy provides a probe of the iron atom in biological samples that is sensitive to changes in oxidation and spin state and the configuration of the ligands around the iron. May et al. [9] grew wheat in an ^{57}Fe-enriched medium and were able to show that the iron that is associated with the phytate in situ in the bran is in the same combination as that in monoferric phytate. For the nuclei Mössbauer spectroscopy is capable of measuring, it deserves further investigation.

2.2. Neutron Activation Analysis

Neutron activation analysis (NAA) involves the production of a radio-active isotope by the capture of neutrons by the nuclei of a stable isotope of an element and subsequent measurement of the radioactivity. Many elements upon irradiation with neutrons give rise to a species of the same atomic number but of one mass unit larger through a neutron-γ reaction. For example, chromium consists of four stable isotopes: ^{50}Cr (4.3%), ^{52}Cr (83.8%), ^{53}Cr (9.6%), and ^{54}Cr (2.4%). Upon irradiation with neutrons, the ^{50}Cr becomes ^{51}Cr, a radioactive isotope having a half-life of 28 days and emitting γ rays with an energy of 0.32 MeV. Naturally, any neutrons absorbed by ^{52}Cr and ^{53}Cr would merely produce another stable isotope (^{53}Cr and ^{54}Cr, respectively). The ^{55}Cr produced from ^{54}Cr is a β-emitter with a half-life of only 3.6 min.

Following irradiation of sample, the induced radioactivity is measured, usually by a semiconductor detector and a multichannel analyzer. This permits resolution of the characteristic radiations produced by the various species formed.

Neutron activation analysis has several advantages over other trace element analysis procedures, in addition to being applicable to stable isotope dilution methods in many cases. Some of these are: (1) very high sensitivity; (2) high specificity because the radio-activity induced is characteristic of the species analyzed (although unresolved overlapping photopeaks do occur); (3) interference from

other elements during the analysis is minimized because following
irradiation the sample's integrity can be threatened only by contamination with radioactive material; and (4) it is capable of
multielement analysis.

Some of the limitations of NAA include: (1) access to a nuclear
reactor and rather extensive counting facilities are required; (2)
many isotopes cannot be measured due to too small a neutron absorption
cross-section, the isotope produced being nonradioactive, having too
long or too short a half-life, or emitting radiation difficult to
measure such as weak β or electron capture. Positron emitters, of
which there are over 50, pose another problem because they all produce
0.511 MeV annihilation radiation.

2.3. Spark Source Mass Spectrometry (SSMS)

The high-voltage spark source is capable of ionizing all elements.
Consequently, in isotope dilution procedures, all elements that can
be separated from the sample and isotopically equilibrated can theoretically be determined at the same time. Moreover, all elements can
be detected with approximately the same sensitivity. These two
factors--multielement capability and uniform sensitivity--are advantages over thermal ionization mass spectrometry (TIMS). Moreover,
the chemical-processing procedures involved as a preliminary to isotopic dilution determination with SSMS are generally simpler than
with TIMS procedures. However, results by isotope dilution TIMS
procedures are more accurate than by SSMS.

The capability of simultaneously determining trace elements was
first investigated at the U.S. National Bureau of Standards (NBS) in
samples by isotopic dilution spark source mass spectrometry (ID-SSMS)
[10-12]. The basic procedure consists of dissolving the sample, isotopically and chemically equilibrating the elements of interest with
known amounts of isotopically enriched isotopes, electrodepositing
the equilibrated isotopes onto gold wires sparking the wires, and
measuring the altered isotopic ratios. Gold was selected because it
is mononuclidic and available in high purity.

ID-SSMS procedures were applied to the characterization and certification of the first two biological Standard Reference Materials (SRMs) issued by the NBS. Cadmium, copper, lead, molybdenum, selenium, silver, thallium, and zinc were determined simultaneously in a freeze-dried bovine liver material which was subsequently issued as SRM 1577. Copper, iron, lead, nickel, selenium, and zinc were determined simultaneously in a dry, powdered orchard leaves material, which was subsequently issued as SRM 1571.

In these procedures, 50-ml quartz Erlenmeyer flasks were used to dissolve and wet-oxidize the sample with nitric and perchloric acids. The procedure was done in two ways. In one, the enriched isotope spikes were added initially to the samples before oxidation, and in the second procedure the samples were oxidized and the spikes were added at the end of the wet-oxidation step. The slight change of procedure was to determine if any elements would be lost before the equilibration step.

In the aforementioned procedures, the trace elements being determined were not appreciably volatile under the experimental conditions employed. However, the volatility of mercury required a special apparatus for wet-oxidizing the orchard leaves sample and equilibrating the mercury with the enriched isotope spike [13]. The apparatus consists of a 250-ml flask, a reflux collector, a 200-mm water-cooled (West) condenser, and a bubbler. This arrangement enables the mercury in the sample to be isotopically diluted with the ^{201}Hg used as the spike without loss. Also, in this procedure ^{198}Hg was used as a "carrier" to assist in electrodepositing the mercury isotopes being measured.

2.4. Thermal Ionization Mass Spectrometry

The thermal ionization mass spectrometry procedure is perhaps the most accurate and precise of all of the stable isotope methods. It is also one of the most time consuming, requires considerable operator skill, and the instrumentation is quite expensive. Perhaps its

most valuable function is as a reference method for the determination
of elements in biological materials by isotope dilution.

A number of scientists at NBS have used this technique with
great success for trace element determinations in biological materials, particularly as part of their SRM certification program ([14-
17], to cite but a few examples). An excellent overview of these
procedures has recently been published by Barnes et al. [18] in discussing the certification of lead concentrations in SRMs by isotope
dilution mass spectrometry (ID-MS).

2.5. Combined Gas Chromatography-Mass Spectrometry (GC/MS)

Another mass spectrometric technique for measuring stable isotopes
of the elements employs a quadrupole mass spectrometer coupled to a
gas chromatograph. The quadrupole instruments do not employ magnets
for ion dispersion and have the advantage of lower cost.

In practice, the analyte in the sample is converted to a stable,
volatile chelate which is then extracted into a suitable solvent and
introduced into the mass spectrometer via the gas chromatograph. This
separates the chelate from the solvent and any interfering species.
Thus, the gas chromatograph serves primarily as a means of sample
introduction.

As an example, let us look at the GC/MS determination of
selenium in biological materials by stable isotope dilution, as
reported recently by Reamer and Veillon [19]. Samples are digested
in a $H_3PO_4/HNO_3/H_2O_2$ mixture and the resulting selenite chelated with
4-nitro-o-phenylenediamine to form the thermally stable nitropiaz-
selenol Se-NPD. This is extracted into chloroform and aliquots introduced into the gas chromatograph. The mass spectrum in the region of
the parent ion, Se-NPD$^+$, is shown in Fig. 1. Here one observes several ion peaks, corresponding to Se-NPD$^+$ containing each of the six
selenium stable isotopes, ^{74}Se, ^{76}Se, ^{77}Se, ^{78}Se, ^{80}Se, and ^{82}Se.
As before, by spiking samples with, say, a known amount of enriched

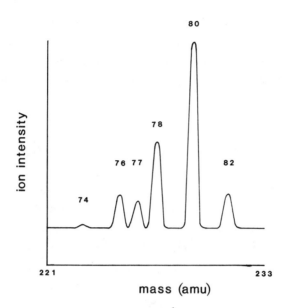

FIG. 1.   Mass spectrum of the Se-NPD$^+$ ion.   Numbers above each peak
refer to the corresponding Se isotope.

^{82}Se and monitoring the ^{80}Se/^{82}Se isotope ratio, one can then calculate the amount of original Se present in the sample.   In GC/MS
methods, a small correction must be made for the amounts of naturally
occurring isotopes in the organic ligand, such as ^{13}C, ^{15}N, ^{18}O, etc.,
but their contributions are readily predictable.   Several other groups
have employed GC/MS methodology for stable isotope dilution and tracer
studies, and these will be described in a later section.

The main advantages of GC/MS methodology over other MS and
nuclear methods are considerably shorter analysis times and less
expensive instrumentation.

The main limitation of GC/MS methods is the availability of
suitable chelating agents.   For several biologically important elements, a suitable chelating agent has not yet been found, but this
does not necessarily mean that one does not exist.

Relative to techniques like TIMS, the measurement precision
attainable with quadrupole instruments is lower, requiring higher

enrichments of an isotope in tracer studies, especially long-term
ones. However, for many applications of tracers and for trace metal
analysis by isotope dilution, the precision attainable is perfectly
adequate. In many cases, the rapid analysis time more than offsets
these limitations.

3. SAMPLE PREPARATION AND CONTAMINATION CONTROL

The main limitation to determining trace elements accurately by
isotopic dilution techniques is the analytical or method blank.
However, by selecting laboratory vessels of inert material, by controlling contamination from the laboratory environment, and especially by limiting and carefully purifying the number of reagents
used, elements in biological systems can be determined at nanogram
per gram concentration levels and below.

Wet oxidation of the biological matrix is frequently the first
step in the isotopic dilution procedure. Although desirable as a
preconcentration or enrichment step, wet oxidation can produce a
large blank depending on the purity of the acids with respect to
elements being determined.

Acid purification by distilling the acids at a temperature
below the boiling point (sub-boiling distillation) has been found
to produce ultrahigh-purity acids [20]. In this procedure, the acid
is vaporized at the surface without boiling and then condensed.

The commercially available sub-boiling still is made of quartz.
Heating of the acid being distilled is done by a pair of infrared
radiators, positioned on both sides of the condenser. These elements,
inside quartz tubes, heat the surface of the liquid and cause it to
evaporate without boiling. This positioning of the heaters also
serves to heat the upper wall of the still above the liquid, which
tends to keep the wall dry and minimize creep of liquid along the
wall between the liquid reservoir and the condensing cold finger.
The condenser is tilted downward toward the distillate outlet to
allow the condenser liquid to flow to the tip above the outlet.

The still is fed by a 3- to 4-kg bottle of ACS reagent grade acid
through a liquid level control which maintains the liquid to just
below the overflow height. All parts of the liquid feed system are
made of Teflon.

As an example of the extent of purification achieved by sub-
boiling distillation, the total concentration of impurity elements,
determined in the purified nitric acid by isotope dilution, spark
source mass spectrometry, was found to be a factor of approximately
100 lower than that of ACS reagent grade acid. The total concentration of the 18 elements determined in the purified nitric acid was
2.3 ng/g.

Similarly for perchloric acid, the total for 17 elements in
the purified acid was 16 ng/g compared with 3400 ng/g for the ACS
reagent grade acid.

Storage of sub-boiling distilled acids is in Teflon-FEP bottles,
which have been thoroughly cleaned initially with reagent acid and
then with purified acid. The bottles are used repeatedly to contain
the same purified acid.

It is advisable to use laboratory ware of either quartz or
Teflon-FEP when determining common elements at the trace level. The
laboratory ware should be heated in acid before using.

Evaporation should be done in a clean environment evaporator
or, if many samples are to be treated, they should be done in a
special class 100 fume hood made of acid-resistant material. Murphy
has reviewed the role of the analytical blank in accurate trace
analysis [21].

4. APPLICATIONS

4.1. Trace Metal Determinations

Isotope dilution mass spectrometric (ID-MS) methods have been developed at the NBS to determine elements accurately in homogeneous biological materials. Samples of these materials are subsequently

issued by NBS as Standard Reference Materials [22]. These SRMs
provide a means of transferring measurement accuracy from NBS to
other laboratories.

Table 1 shows the nine NBS biological SRMs that are available
and lists the concentrations of the certified elements. In addition,
the Certificate of Analysis shows the estimated uncertainties of the
concentrations based on the methods and possible systematic errors.
The certificate also gives uncertified concentrations of elements
determined by one method.

In addition to analyses by ID-MS procedures, analyses are also
made by other independent analytical methods. Concordant results by
two or more independent methods are generally required to provide a
certified concentration value for an element. However, the inherent
accuracy of analytical determinations of elements, such as lead and
calcium, by ID-TIMS is so high that certified concentrations may be
based on these determinations alone. These methods have been called
definitive methods.

4.2. Tracer Studies

One of the most important applications of stable isotopes is their
use in tracer studies. In particular, they afford a means of investigating trace element metabolism in humans. This is a very important
feature because in many cases the use of radiotracers for these purposes is contraindicated.

For those elements having more than two stable isotopes, it is
possible to use one as a tracer or metabolic tag and spike samples
with another for isotope dilution measurements. In this way, one
can quantitatively identify the metabolic tag, the unenriched analyte
present with it, and the total amount of analyte element present,
which is simply the sum of these. In cases where only two stable
isotopes of an element exist or only two can be measured, the total
amount of analyte present would normally have to be determined by an

alternate method. However, there is a way around this two-isotope limitation (see below). Details of how these measurements are performed have been described in detail elsewhere [23].

Several groups in the United States have been active in recent years in employing stable isotopes of various elements in tracer studies involving trace element metabolism. One particularly active group in this regard has recently reported on such studies as using stable isotopes to measure bioavailability by fecal monitoring [24], measurements of ^{70}Zn enrichment in human plasma samples [25], and metabolic kinetics following ingestion of ^{74}Se [26], to cite but a few examples of this prolific group's work. The majority of these studies employed NAA for the stable isotope measurements.

Another group employing NAA studied the utilization of iron, using ^{58}Fe as the stable isotope tracer [27]. They were able to demonstrate in this preliminary study that sufficient precision and accuracy could be achieved, even considering the relatively high levels of iron in the body.

In an elegant study, Harvey employed stable isotopes and SSMS to study the uptake of copper by fish from seawater and food [28]. Copper consists of only two isotopes, ^{63}Cu and ^{65}Cu, and no suitably long-lived radioisotopes exist. It was mentioned above that a tracer method that also yields total analyte concentration usually requires that the element have at least three stable isotopes. Harvey got around this cleverly in his study in the following way: ^{65}Cu was used as the tracer in the uptake experiments and ^{63}Cu was used to spike the samples (internal standard). Measuring the isotope ratios before and after the samples were spiked, it is possible to calculate both the amount of naturally occurring Cu present and the excess ^{65}Cu (which is a measure of the tracer taken up by the fish during the experiment).

Another active group in California recently measured zinc and iron absorption in elderly men using the stable isotopes ^{70}Zn and ^{58}Fe [29]. They employed TIMS, and found mean apparent absorptions of 17.3 and 7.9% for Zn and Fe, respectively. These values agree

TABLE 1

Composition of NBS Biological Standard Reference Materials*

Element	Oyster tissue 1566	Bovine liver 1577a	Wheat flour 1567	Rice flour 1568	Brewer's yeast 1569	Orchard leaves 1571	Citrus leaves 1572	Tomato leaves 1573	Pine needles 1575
Aluminum	--	--	--	--	--	--	92	--	545
Antimony	--	--	--	--	--	2.9	--	--	--
Arsenic	13.4	0.047	--	0.41	--	10	3.1	0.27	0.21
Barium	--	--	--	--	--	--	21	--	--
Beryllium	--	--	--	--	--	0.027	--	--	--
Boron	--	--	--	--	--	33	--	--	--
Cadmium	3.5	0.44	0.032	0.029	--	0.11	0.03	--	--
Calcium	0.15%	120	0.019%	0.014%	--	2.09%	3.15%	3.00%	0.41%
Chlorine	--	0.28%	--	--	--	--	--	--	--
Chromium	0.69	--	--	--	2.12	2.6	0.8	4.5	2.6
Cobalt	--	0.21	--	0.02	--	--	--	--	--
Copper	63.0	158	2.0	2.2	--	12	16.5	11	3.0
Iron	195	194	18.3	8.7	--	300	90	690	200

Element									
Lead	0.48	0.135	0.020	0.045	--	45	13.3	6.3	10.8
Magnesium	0.128%	600	--	--	--	0.62%	0.58%	--	--
Manganese	17.5	9.9	8.5	20.1	--	91	23	238	675
Mercury	0.057	0.004	0.001	0.0060	--	0.155	0.08	--	0.15
Molybdenum	--	3.5	--	--	--	0.3	0.17	--	--
Nickel	1.03	--	--	--	--	1.3	0.6	--	--
Nitrogen	--	--	--	--	--	2.76%	--	--	--
Phosphorus	--	1.11%	--	--	--	0.21%	0.13%	0.34%	0.12%
Potassium	0.969%	0.9967%	0.136%	0.112%	--	1.47%	1.82%	4.46%	0.37%
Rubidium	4.45	12.5	--	--	--	12	4.84	16.5	11.7
Selenium	2.1	0.71	1.1	0.4	--	0.08	--	--	--
Silver	0.89	0.04	--	--	--	--	--	--	--
Sodium	0.51%	0.243%	8.0	6.4	--	82	160	--	--
Strontium	10.36	0.138	--	--	--	37	100	44.9	4.8
Sulfur	--	0.78%	--	--	--	--	0.407%	--	--
Thorium	--	--	--	--	--	0.064	--	0.17	0.037
Uranium	0.116	0.00071	--	--	--	0.029	--	0.061	0.020
Zinc	852	123	10.6	19.4	--	25	29	62	--

*Content in µg/g (or, where noted, wt %).

very well with those obtained in earlier studies on the same subjects using radioisotopes.

Yergey and co-workers recently reported on measuring stable isotopes of calcium in urine using TIMS [30]. One really unique aspect of their work is that they cleverly converted a conventional quadrupole instrument to thermal ionization by a modification of the original solids probe on the instrument. The principal advantages demonstrated for this system compared with other TIMS systems include rapid analysis time, acceptable accuracy for tracer studies, and relatively inexpensive instrumentation.

The remaining tracer studies to be discussed all involve measurement of the stable isotopes by mass spectrometry employing volatile chelates.

Schwartz and Giesecke investigated the use of ^{26}Mg as a tracer using a volatile chelate introduced into a quadrupole MS via a solids probe [31]. Subjects were fed both ^{26}Mg and the short-lived radioisotope ^{28}Mg; urine, feces, and plasma were monitored by γ counting, neutron activation, and mass spectrometry. They judged their method superior to NAA for Mg tracer studies.

Using isotope dilution GC/MS, Hachey and co-workers investigated a series of Mg, Ca, Cr, Fe, Ni, Cu, Cd, and Zn chelates [32]. Using conventional GC/MS instrumentation and both electron and chemical ionization, isotopic abundances were determined at accuracies between 0.01 and 1.19 at.% and precisions between 0.01 and 0.27 at.% (RSD ± 0.07-10.26%).

Using a solids probe and quadrupole MS, Johnson investigated mineral metabolism in human subjects employing ^{54}Fe, ^{57}Fe, ^{67}Zn, ^{70}Zn, and ^{65}Cu [33]. The metals were chelated with tetraphenylporphyrin, and isotope ratios could be measured in samples containing as little as 10 ng of metal in plasma, red cells, urine, feces, and sweat. Altered isotopic ratios could be detected for 3-4 months after the tracer dose for zinc and for nearly 6 months for iron.

Reamer and Veillon have described in detail a double isotope dilution method for using stable selenium isotopes in metabolic tracer studies by GC/MS [23]. This method is rapid and precise, and avoids the use of perchloric acid in sample preparation. The methodology was subsequently employed in monitoring the intrinsic labeling of chicken products with ^{76}Se [34]. These endogenously labeled products were subsequently used in a controlled metabolic study to measure Se uptake in pregnant women [35]. Excellent results were obtained, demonstrating the suitability of stable isotopes for tracer studies in a situation where radioisotopes could not be used, and demonstrating a clear superiority over conventional balance methods. Absorption of the ^{76}Se tracer was found to be very high (85%) for all three groups studied: nonpregnant (NP), early pregnant (EP), and late pregnant (LP). However, the Se retained by the three groups was LP > EP > NP, presumably due to the fetus.

REFERENCES

1. D. M. Bier, *Nutr. Rev.*, *40*, 129 (1982).

2. J. A. Goleb, *Anal. Chim. Acta*, *34*, 135 (1966).

3. W. H. Brimhall, *Anal. Chem.*, *41*, 1349 (1969).

4. K. R. Osborn and H. E. Gunning, *J. Opt. Soc. Am.*, *45*, 552 (1955).

5. C. Veillon and J. Y. Park, *Anal. Chem.*, *44*, 1473 (1972).

6. J. A. Goleb and Y. Yokoyama, *Anal. Chim. Acta*, *30*, 213 (1964).

7. F. E. Lichte and R. K. Skogerboe, *Appl. Spectrosc.*, *28*, 354 (1974).

8. J. A. Goleb, *Anal. Chim. Acta*, *36*, 130 (1966).

9. L. May, E. R. Morris, and R. Ellis, *J. Agric. Food Chem.*, *28*, 1004 (1980).

10. R. Alvarez, P. J. Paulsen, and D. E. Kelleher, *Anal. Chem.*, *41*, 955 (1969).

11. P. J. Paulsen, R. Alvarez, and D. E. Kelleher, *Spectrochim. Acta*, *24B*, 535 (1969).

12. P. J. Paulsen, R. Alvarez, and C. W. Mueller, *Anal. Chem.*, *42*, 674 (1970).

13. R. Alvarez, *Anal. Chim. Acta*, *73*, 33 (1970).

14. L. J. Moore, L. A. Machlan, W. R. Shields, and E. L. Garner, *Anal. Chem.*, *46*, 1082 (1974).

15. E. L. Garner, L. A. Machlan, J. W. Gramlich, L. J. Moore, T. J. Murphy, and I. L. Barnes, *Nat. Bur. Stds. Spec. Tech. Publ.*, *442*, 951 (1976).

16. L. P. Dunstan and E. L. Garner, in *Trace Substances in Environmental Health--XI* (D. D. Hemphill, ed.), Univ. of Missouri, Columbia, 1977, p. 334.

17. E. L. Garner and L. P. Dunstan, *Adv. Mass Spectrom.*, *74*, 481 (1978).

18. I. L. Barnes, T. J. Murphy, and E. A. I. Michiels, *J. Assoc. Off. Anal. Chem.*, *65*, 953 (1982).

19. D. C. Reamer and C. Veillon, *Anal. Chem.*, *53*, 2166 (1981).

20. E. C. Kuehner, R. Alvarez, P. J. Paulsen, and T. J. Murphy, *Anal. Chem.*, *44*, 2050 (1972).

21. T. J. Murphy, *Nat. Bur. Stds. Spec. Tech. Publ.*, *422*, 509 (1976).

22. *Catalog of Standard Reference Materials*, *Nat. Bur. Stds. Spec. Tech. Publ.*, *260*, Nov. 1981. (Available from the Superintendent of Documents, U.S. Government Printing Office, Washington, D.C., 20402.)

23. D. C. Reamer and C. Veillon, *J. Nutr.*, *113*, 786 (1983).

24. M. Janghorbani and V. R. Young, *Am. J. Clin. Nutr.*, *33*, 2021 (1980).

25. M. Janghorbani, V. R. Young, J. W. Gramlich, and L. A. Machlan, *Clin. Chim. Acta*, *114*, 163 (1981).

26. M. Janghorbani, M. J. Christensen, A. Nahapetian, and V. R. Young, *Am. J. Clin. Nutr.*, *35*, 647 (1982).

27. J. J. Carni, W. D. James, S. R. Koirtyohann, and E. R. Morris, *Anal. Chem.*, *52*, 216 (1980).

28. B. R. Harvey, *Anal. Chem.*, *50*, 1866 (1978).

29. J. R. Turnlund, M. C. Michel, W. R. Keyes, J. C. King, and S. Margen, *Am. J. Clin. Nutr.*, *35*,1033 (1982).

30. A. L. Yergey, N. E. Vieira, and J. W. Hansen, *Anal. Chem.*, *52*, 1811 (1980).

31. R. Schwartz and C. C. Giesecke, *Clin. Chim. Acta*, *97*, 1 (1979).

32. D. L. Hachey, J. C. Blais, and P. D. Klein, *Anal. Chem.*, *52*, 1131 (1980).

33. P. E. Johnson, *J. Nutr.*, *112*, 1414 (1982).

34. C. A. Swanson, D. C. Reamer, C. Veillon, and O. A. Levander, *J. Nutr.*, *113*, 793 (1983).

35. C. A. Swanson, J. C. King, O. A. Levander, D. C. Reamer, and C. Veillon, *Am. J. Clin. Nutr.* (in press).

Chapter 6

TRACE ELEMENTS IN CLINICAL CHEMISTRY DETERMINED BY NEUTRON ACTIVATION ANALYSIS

Kaj Heydorn
Isotope Division
Risø National Laboratory
Roskilde, Denmark

1. INTRODUCTION

The activation of natural elements by bombardment with α particles
was discovered by the Joliot-Curies in 1934, and soon after Hevesy
and Hilde Levi [1] applied such artificial radioactivity, induced
by exposure to neutrons, to the analysis of rare earth elements.

This *instrumental neutron activation analysis* (INAA) remained
a curiosity until the advent of the nuclear reactor, which made possible the exposure to neutron flux densities millions of times higher
than those previously available. The resulting increase in sensitivity for all elements is extremely important for determining the very
low concentrations of many elements of interest in biological material.

1.1. Principles of Neutron Activation Analysis

In 1949 the *comprehensive* neutron activation analysis was described
by Boyd [2] in the form most suitable for the determination of trace
elements in samples of clinical interest. The sample is irradiated
without any preceding treatment, and after activation it is totally
decomposed and brought into solution in the presence of a carrier of
the element to be determined. Radiochemical separation serves to
eliminate interference from other radionuclides in the sample before
measurement, and the determination of recovery of the added carrier
makes it possible to correct for possible losses during the processing of the sample.

Used properly, this *radiochemical neutron activation analysis*
(RNAA) qualifies as a reference method, provided a weighed quantity
of determinand is used as a comparator standard. RNAA is often used
in the certification of reference materials, but only for the determination of total element content, regardless of chemical composition.

Let a sample contain N_A atoms of an element to be determined,
A, with a relative atomic mass A_r. When exposed to a neutron flux
density ϕ, a small fraction of A is transformed into a radioactive
isotope B:

$$-\frac{dN_A}{dt} = \sigma\phi N_A \tag{1}$$

where the factor σ is the activation cross-section. The formation of B according to Eq. (1) is accompanied by its decay, and the number of atoms N_B is determined by

$$\frac{dN_B}{dt} = \sigma\phi N_A - \lambda N_B \tag{2}$$

where λ is the decay constant. The activity of B after activation for a time t_i is found by integration of Eq. (2):

$$\lambda N_B = N_A \sigma\phi(1 - e^{-\lambda t_i}) \tag{3}$$

Radiochemical separation takes place during a decay time t_d, and the number of atoms of B decaying during the succeeding measurement time t_m becomes

$$N_B[e^{-\lambda t_d} - e^{-\lambda(t_d + t_m)}] \tag{4}$$

or, after rearrangement and combination with Eq. (3),

$$\frac{N_A \sigma\phi}{\lambda}(1 - e^{-\lambda t_i})(1 - e^{-\lambda t_m})e^{-\lambda t_d} \tag{5}$$

For N_A equal to Avogadro's number, expression (5) gives the sensitivity as the number of disintegrations per A_r gram of the element A.

2. SPECIAL ADVANTAGES FOR ULTRATRACE ANALYSIS

In the period up to the beginning of the sixties no other analytical method could match the sensitivity of NAA as well as the detection limit achieved for many trace elements by RNAA. During the last decade, however, several other analytical methods have been developed with sufficient sensitivity to permit the determination of many trace elements for clinical purposes.

2.1. Blank Value

At the concentrations of less than 10 μg/kg, often referred to as
the *ultratrace level* [3], high sensitivity is not the only performance characteristic to be considered. The most serious problems are
often associated with the uncertainty of the blank [4]. Here NAA
has the unique characteristic that impurities in the reagents used
in the radiochemical separation, as well as any other contamination
of the sample after activation, does not affect the result of the
measurement of the corresponding radioactive isotope. Contamination
therefore need be considered only in connection with the actual
sampling process and preparation for neutron activation, and for
many elements blank values can be reduced to insignificance [4].

2.2. Sensitivity for Different Elements

When the blank can be neglected, the detection limit that can be
achieved after radiochemical separation is determined by the sensitivity of the element, calculated in accordance with expression (5).
Such calculations have been carried out by Guinn and Hoste [5] for
an irradiation time t_i of 5 hr, a counting time t_m of 100 min, and
zero decay time. With a flux density ϕ of 10^{17} neutrons/$(m^2 s)$ and
using a typical germanium-lithium detector for the measurement of
γ rays, they reported detection limits corresponding to 30 counts
in the major photopeak for 68 elements.

Two thirds of these--or 42 elements--have detection limits of
1 ng or less and may therefore be determined at the ultratrace level
in samples of 0.1 g or more. In Table 1 these elements are grouped
according to increasing sensitivity.

Among the elements listed in the table, only chlorine and
sodium occur in biomedical samples in concentrations above the trace
level of 100 mg/kg. Immediately after irradiation ^{38}Cl is the domi-
nant activity, but after a few hours ^{24}Na accounts for an activity
level such that other trace elements can be determined only after a

TABLE 1

Elements with Calculated Limits of NAA Detection
of <1 ng in the Absence of Interfering Activity
After 5 hr Irradiation at 10^{17} neutrons/(m^2 s)

Element	Indicator nuclide	Half-life	γ energy (keV)
Detection limit < 1 ng			
Barium	^{139}Ba	82.9 min	166
Chlorine	^{38}Cl	37.3 min	1642
Gadolinium	^{159}Gd	18 hr	364
Germanium	^{75}Ge	82 min	265
Hafnium	^{180}Hf^m	5.5 hr	215
Neodymium	^{149}Nd	1.8 hr	211
Scandium	^{46}Sc	83.9 days	889
Selenium	^{77}Se^m	17.5 sec	162
Strontium	^{87}Sr^m	2.83 hr	389
Tantalum	^{182}Ta	115 days	68
Tellurium	^{131}Te	24.8 min	160
Terbium	^{160}Tb	72.1 days	299
Thorium	^{233}Pa	27 days	310
Thulium	^{170}Tm	134 days	84
Detection limit < 0.1 ng			
Antimony	^{122}Sb	2.8 days	564
Arsenic	^{76}As	26.4 hr	559
Bromine	^{82}Br	35.3 hr	554
Caesium	^{134}Cs^m	2.89 hr	127
Cobalt	^{60}Co^m	10.5 min	59
Copper	^{64}Cu	12.8 hr	511
Erbium	^{171}Er	7.5 hr	308
Gallium	^{72}Ga	14.1 hr	834
Iodine	^{128}I	25.0 min	443
Lanthanum	^{140}La	40.2 hr	487
Mercury	^{197}Hg	65.0 hr	69-78

TABLE 1 (Continued)

Element	Indicator nuclide	Half-life	γ energy (keV)
	Detection limit < 0.1 ng		
Palladium	^{109}Pd	13.5 hr	88
Ruthenium	^{105}Ru	4.4 hr	724
Sodium	^{24}Na	15.0 hr	1368
Tungsten	^{187}W	23.9 hr	72
Vanadium	^{52}V	3.75 min	1434
Ytterbium	^{177}Yb	1.9 hr	150
	Detection limit < 0.01 ng		
Gold	^{198}Au	2.70 days	412
Iridium	^{194}Ir	17.4 hr	328
Lutetium	^{176}Lu^m	3.7 hr	88
Manganese	^{56}Mn	2.58 hr	847
Rhenium	^{188}Re	16.7 hr	155
Rhodium	^{104}Rh^m	4.4 min	51
Samarium	^{153}Sm	46.8 hr	103
	Detection limit < 0.001 ng		
Dysprosium	^{165}Dy	139 min	95
Europium	^{152}Eu$^m1$	9.3 hr	122
Holmium	^{166}Ho	26.9 hr	81
Indium	^{116}In$^m1$	54.0 min	417

Source: Ref. 5.

radiochemical separation. Once the ^{24}Na activity has decayed to a negligible level, some trace elements may be determined without radiochemical separation [6], including iron, zinc, and cobalt, discussed in Chap. 9.

3. PRECISION AND ACCURACY

Experimental errors affecting analytical results may be divided into
random and systematic contributions. The random part is conveniently
characterized by a standard deviation and referred to by the term
precision, while *accuracy* refers to the systematic part, characterized by a bias [7]. Precision is therefore studied by replicate
analyses of typical samples, and accuracy is checked by the analysis
of reference materials with known concentration of determinand.

Analytical methods based on the counting of radionuclides are
subject to a special source of random variation, called *counting
statistics.* The exponential distribution of intervals between successive decays gives rise to a counting variance equal to the mean
number of counts. This characteristic of the Poisson distribution
makes it possible to estimate the standard deviation of a single
measurement, without any replication. For INAA, as well as for RNAA
at the ultratrace level, counting statistics is a significant part
of the total random error.

3.1. A Priori Precision

The uncertainty of an analytical result is determined by the precision of the analytical method [8], which again represents the overall
effect of the procedures and techniques applied. It is therefore
possible to predict the standard deviation $\hat{\sigma}$ of a result y before the
analysis has been carried out, and this is conveniently referred to
as the *a priori precision.*

In neutron activation analysis the a priori precision must be
compounded with the contribution from the counting process, and in
INAA the standard deviation is determined by the possible uncertainty
of the quantity of sample, the uncertainty in the activation in relation to the comparator standard, together with the counting statistics. In RNAA we have additional sources of variation similar to

those encountered in other methods with a separation step; if the radiochemical yield is determined, most other sources of variation are replaced by the uncertainty of the yield determination.

During separation the carrier addition maintains a constant concentration of determinand, and therefore NAA, unlike other methods of analysis, is characterized by an a priori precision which is independent of the concentration.

3.2. Statistical Control

When all sources of random variation have been detected and accounted for the predicted standard deviations $\hat{\sigma}_i$ of analytical results y_i accurately account for the observed variability of replicate determinations, and the method is said to be in a state of *statistical control*. This is tested by the statistic:

$$T = \sum_1^n \frac{(y_i - \hat{\mu})^2}{\hat{\sigma}_i^2} \tag{6}$$

where $\hat{\mu}$ is the weighted mean of n replicates.

When the replicates are in statistical control, this statistic is approximately described by a χ^2 distribution with n - 1 degrees of freedom.

In NAA the estimated standard deviation must of course include the counting statistics, and by replicate analysis of reference materials at both high and low levels of determinand it is possible to verify both the a priori precision and the counting statistics.

Only at that stage is it possible to demonstrate the accuracy of the method by the analysis of certified reference materials.

4. ANALYSIS OF PRECISION

When the analytical method has been brought into a state of statistical control by the analysis of various reference materials, the statistic T of Eq. (6) may be used as a continuous quality control [9]. In radiochemical neutron activation analysis the absence of concentration and matrix effects extends the applicability of Eq. (6) to the ultratrace level, even if the statistical control has been demonstrated only at the trace level. This application of the T statistic is sometimes referred to as the *analysis of precision* [8].

4.1. Sampling Errors

The use of the analysis of precision on the results from replicate analysis of actual biomedical samples, rather than reference materials, often leads to the detection of additional sources of variability by high values of T.

Lack of homogeneity may be responsible for considerable variation among replicate samples and must be taken into account. This is conveniently done by means of the sampling constant K_s according to Ingamells and Switzer [10]:

$$K_s = s^2 m \tag{7}$$

where m is the mass of the sample and s the percent relative standard deviation characterizing the degree of homogeneity.

Although originally introduced for geochemical samples, this concept has been applied to biological material as well [11]. K_s may be determined experimentally from the analysis of replicate samples by analytical methods in statistical control [12,13]; for heterogeneous organs like the kidney or placenta, sampling constants of several hundred grams have been found.

4.2. Contamination Problems

In biomedical samples of a high degree of homogeneity, such as blood or urine, high values of T in the analysis of precision must be attributed to other problems. At the ultratrace level contamination has been found to lead to considerable divergence, and attempts to eliminate airborne contamination resulted in a significant reduction of the value of T [12].

Used in this way the analysis of precision serves to detect systematic errors from contamination by means of the statistical test in Eq. (6) for random sources of variation [3].

5.   EXAMPLES (Not Mentioned in Chaps. 9 and 13)

In Chaps. 9 and 13 the use of NAA for the determination of the elements manganese, iron, cobalt, copper, zinc, and gold is discussed in relation to other methods of analysis.

Among the elements listed in Table 1 many others have attracted attention, although only a few have known biological functions, such as iodine.

5.1. Selenium

Selenium is an essential element with normal concentrations in soft tissue from 50 to 500 μg/kg [14], and it therefore represents a borderline case between trace and ultratrace elements. It may be determined by INAA, as well as by RNAA, using several different indicator nuclides; for urine and blood fluorometry has been used with comparable results.

Several groups use RNAA with ^{75}Se as an indicator for the determination of selenium in the liver, kidney, and other organs [15-18], whereas Larsen et al. [19] rely on ^{81}Se^m for the analysis of brain tissue.

Kasperek et al. [20] used INAA with ^{75}Se as indicator to search for possible differences between the concentration of selenium in serum and plasma from the same patient. Heydorn et al. [21] used RNAA with ^{81}Se^m to detect changes in the serum selenium level of a person who was given a meal after 24 hr of fasting. Results shown in Fig. 1 illustrate the considerable short-term stability of selenium in blood serum, while the analysis of precision demonstrated significant differences from one individual to another.

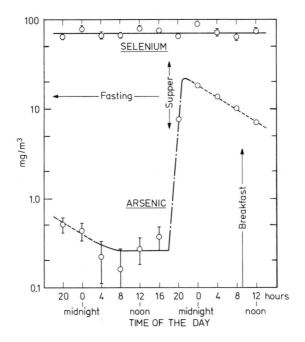

FIG. 1. Concentrations of selenium and arsenic in serum from a healthy young woman during 24 hr of fasting followed by a supper containing a fillet of plaice [21].

Considerable geographic differences in selenium intake are known to occur and are reflected in the concentration of selenium in the blood. For rapid screening of many samples for excess or deficiency, the short-lived $^{77}Se^m$ is a useful indicator for INAA, provided the samples are dialyzed and lyophilized before irradiation [22].

5.2. Arsenic

Although probably not an essential element, arsenic at the trace level has always attracted attention, presumably because of its well-known acute toxicity at higher levels. Average concentrations in healthy individuals are at the ultratrace level, where RNAA is

the only reliable method [14]; for the measurement of toxic levels
atomic absorption spectrometry and other methods may well be used.

The only indicator isotope is [76]As with a half-life of 26.4 hr,
and most, but not all, separation methods are based on volatilization
of the hydride or the halides [15,16,18]. Arsenic affects the metabolism of selenium compounds, and simultaneous determination of both
elements is therefore of interest [15,16,19].

Lack of homeostatic control results in wide variations among
individuals, and the levels of arsenic in blood were found to be
closely approximated by a log-normal rather than a normal distribution [23]. Even then significant geographic differences have been
established, probably reflecting different dietary intakes of arsenic-
containing foods. Even in a single individual the concentration of
arsenic in serum was found to rise by approximately two orders of
magnitude over a period of a few hours following the consumption of
plaice with a few parts per million of arsenic [21]. This is shown
in Fig. 1 for the same person analysed with respect to selenium and
demonstrates the remarkable differences in short-term stability of
these two elements.

5.3. Vanadium

Vanadium has been shown to have an essential function in some animals
and is therefore likely to be essential also to humans. Concentrations in biological material are at the ultratrace level and near the
limit of detection for nearly all analytical methods, including neutron activation analysis.

The indicator isotope is [52]V with a half-life of 3.75 min, and
considerable loss of sensitivity therefore takes place during a complete radiochemical separation. Attempts to apply preconcentration
followed by INAA to biomedical samples led to a limit of detection
of 7 ng [24]; this is satisfactory for plant material [25], but is
just barely sufficient for the analysis of liver and lungs [15,24,
26].

Concentrations in human blood have been determined by Cornelis et al. [26-28] by radiochemical separation of irradiated samples that had been ashed at 450°C before activation. In that case the duration of the separation was less than two half-lives, and the limit of detection was improved by more than an order of magnitude.

The majority of results for vanadium in human serum were in the range of 0.01 to 0.09 mg/m^3, and these are the lowest concentrations hitherto reported for a potentially essential element.

6. CONCLUSION

In the past, neutron activation analysis has been used for many bio-medical studies because of its high sensitivity for a number of trace elements. For several of these elements methods based on atomic absorption spectrometry are now available with sufficient sensitivity and are more suited to routine applications.

At the ultratrace level high sensitivity is not the only criterion, and other characteristics of RNAA assume an even greater role. The absence of reagent blanks and the negligible matrix dependence are of paramount importance at these concentrations. However, the independence of the a priori precision with respect to the concentration of the determinand is that unique characteristic which makes possible the extension of statistical control throughout the analytical range and down to the limit of detection. Neutron activation analysis will therefore continue to be used as a reference method for investigations at the ultratrace level of concentration.

ABBREVIATIONS AND DEFINITIONS

Analytical Terms

Accuracy. The closeness of agreement between the true value and the mean result which would be obtained by applying the experimental procedure a very large number of times.

The smaller the systematic part of the experimental errors which affect the results, the more accurate is the procedure [7].

Activation analysis

NAA = neutron activation analysis

INAA = instrumental neutron activation analysis

RNAA = neutron activation analysis with radiochemical separation

Carrier. An element added to the sample after activation in order to control losses of determinand during radiochemical separation.

Determinand. The element or chemical species to be determined.

Sensitivity. In NAA the number of counts recorded per unit weight of an element under prescribed conditions.

Trace. A concentration of the determinand that is lower than 0.01% or 100 mg/kg.

Ultratrace. A concentration of the determinand that is lower than 0.01 mg/kg or equivalent.

Statistical Concepts

Precision. The closeness of agreement between the results obtained by applying the experimental procedure several times under prescribed conditions.

The smaller the random part of the experimental errors which affect the results, the more precise is the procedure [7].

Analysis of precision. The detection of unexpected sources of variability by comparison of estimated and observed variation of analytical results [8].

A priori precision. A set of instructions for the a priori estimation of the standard deviation of a single analytical result [8].

Statistical control. Full agreement between the estimated and observed variability of analytical results.

REFERENCES

1. G. Hevesy and H. Levi, *K. Dan. Vidensk. Selsk., Mat.-Fys. Medd.,* *14*(5), 34 (1936).

2. G. E. Boyd, *Anal. Chem., 21,* 335 (1949).

3. K. Heydorn, in *Accuracy in Trace Analysis* (P. D. LaFleur, ed.), National Bureau of Standards, Washington, D.C., 1976, p. 127ff.

4. K. Heydorn and E. Damsgaard, *Talanta, 29,* 1019 (1982).

5. V. P. Guinn and J. Hoste, in *Elemental Analysis of Biological Materials* (Technical Report Series No. 197), International Atomic Energy Agency, Vienna, 1980, p. 105ff.

6. R. M. Parr, in *Nondestructive Activation Analysis* (S. Amiel and M. Mantel, eds.), Elsevier/North Holland, Amsterdam, 1981, p. 139ff.

7. International Standard ISO 3534, Statistics--Vocabulary and Symbols, International Organization for Standardization, 1977.

8. K. Heydorn and K. Nørgaard, *Talanta, 20,* 835 (1973).

9. K. Heydorn, in *Development of Nuclear-Based Techniques for the Measurement, Detection and Control of Environmental Pollutants* (Proceedings of a Symposium), International Atomic Energy Agency, Vienna, 1976, p. 61ff.

10. C. O. Ingamells and P. Switzer, *Talanta, 20,* 547 (1973).

11. E. Damsgaard, K. Østergaard, and K. Heydorn, *J. Radioanal. Chem., 70,* 67 (1982).

12. K. Heydorn, *Aspects of Precision and Accuracy in Neutron Activation Analysis,* Risø National Laboratory, Roskilde, 1980, Chap. 5.

13. K. Heydorn, *Neutron Activation Analysis for Clinical Trace Element Research,* Vol. 1, CRC Press, Florida, 1983, Chap. 2.

14. K. Heydorn, in Chap. 6 of Vol. 2 of Ref. 13.

15. P. Lievens, J. Versieck, R. Cornelis, and J. Hoste, *J. Radioanal. Chem., 37,* 483 (1977).

16. J. J. M. deGoeij, K. J. Volkers, and P. S. Tjioe, *Anal. Chim. Acta, 109,* 139 (1979).

17. L.-O. Plantin, in *Nuclear Activation Techniques in the Life Sciences* (Proceedings of a Symposium), International Atomic Energy Agency, Vienna, 1979, p. 321ff.

18. D. Brune, G. F. Nordberg, P. O. Wester, and B. Bivered, p. 643ff of Ref. 17.

19. N. A. Larsen, H. Pakkenberg, E. Damsgaard, and K. Heydorn, *J. Neurol. Sci., 42,* 407 (1979).

20. K. Kasperek, J. Kiem, G. V. Iyengar, and L. E. Feinendegen, *Sci. Total Environ., 17,* 133 (1981).

21. K. Heydorn, E. Damsgaard, N. A. Larsen, and B. Nielsen, in *Nuclear Activation Techniques in the Life Sciences* (Proceedings of a Symposium), International Atomic Energy Agency, Vienna, 1979, p. 129ff.

22. J. R. Vogt, A. Abu-Samra, D. McKown, J. S. Morris, W. D. James, J. Carni, and C. Graham, in *Trace Element Analytical Chemistry in Medicine and Biology* (P. Brätter and P. Schramel, eds.), de Gruyter, Berlin, 1980, p. 447ff.

23. K. Heydorn, *Clin. Chim. Acta, 28,* 349 (1970).

24. E. Damsgaard, K. Heydorn, and B. Rietz, in *Nuclear Activation Techniques in the Life Sciences* (Proceedings of a Symposium), International Atomic Energy Agency, Vienna, 1972, p. 119ff.

25. K. Heydorn and E. Damsgaard, *J. Radioanal. Chem., 69,* 131 (1982).

26. R. Cornelis, L. Mees, J. Hoste, J. Ryckebusch, J. Versieck, and F. Barbier, in *Nuclear Activation Techniques in the Life Sciences* (Proceedings of a Symposium), International Atomic Energy Agency, Vienna, 1979, p. 165ff.

27. R. Cornelis, J. Versieck, L. Mees, J. Hoste, and F. Barbier, *J. Radioanal. Chem., 55,* 35 (1980).

28. R. Cornelis, J. Versieck, L. Mees, J. Hoste, and F. Barbier, *Biol. Trace Element Res., 3,* 257 (1981).

Chapter 7

DETERMINATION OF LITHIUM, SODIUM, AND POTASSIUM IONS IN CLINICAL CHEMISTRY

Adam Uldall
Department of Clinical Chemistry
University of Copenhagen, Herlev Hospital
Herlev, Denmark

and

Arne Jensen
Chemistry Department AD
Royal Danish School of Pharmacy
Copenhagen, Denmark

1. CLINICAL CHEMICAL ASPECTS

1.1. Physiological Chemistry

Sodium ion and potassium ion are essential for the normal functioning of the organism. The whole-body amount of sodium ion and potassium ion are approximately 4 and 3.5 mol, respectively, in adult humans. Nearly 1 mol of sodium is incorporated as nondiffusible salts in the skeleton. The residual amount of sodium and nearly all potassium are diffusible. The concentration of sodium ion in the extracellular fluid is near 140 mmol/liter, whereas it is approximately 10 mmol/ liter inside the cells. There is, however, considerable deviation between the type of cells. The corresponding values for potassium ion are 4 and 140 mmol/liter, respectively. It requires energy for the cell to maintain these concentration differences; Na^+ is transported actively; K^+ and water are transported by diffusion. Na^+ and K^+ are absorbed from the gastrointestinal tract nearly completely and are mainly excreted through the kidney. The elimination is regulated by aldosterone and corticoids, and is influenced by the acid-base regulation. The normal urinary secretion depends on the amount given orally. The concentrations of Na^+ and K^+ in plasma from most of the healthy adults (95%) are in the intervals 137-147 and 3.6-5.0 mmol/liter, respectively [1].

The total protein concentration in human plasma is approximately 70 g/liter. As the protein is dissolved colloidally it has a certain volume. The concentration of water in plasma is near 53 mol/liter and is somewhat less than in pure water (55 mol/liter). As Na^+ and K^+ are dissolved in the aqueous phase, the equilibrium between plasma and the cells depends on the activity of the ions in the noncolloidal phase called the plasma water. The concentration of Na^+ and K^+ in the plasma water seems more easy to interpret than that in plasma,

due to the independence of colloidal proteins and lipids. Values
based on whole plasma are generally used for analytical technical
reasons.

In practice the plasma specimens may be collected in tubes
after venipuncture applying no or minor stasis. The tubes are
coated with ammonium heparinate, the purpose being to inhibit blood
clotting. Within half an hour the plasma should be isolated by
centrifugation and decantation in order to minimize exchange of ions
between the plasma and the cells. The blood is, alternatively, collected in uncoated glass tubes, and the blood will clot within 1 hr.
The fibrinogen-free plasma, called serum, can be isolated, and analyzed after centrifugation. Some authors claim to find slightly
higher concentration of potassium ion in serum than in plasma, but
this is not agreed generally.

1.2. Pathophysiological Chemistry and Pharmacological Background

Elevated concentration of *sodium ion* in plasma is most frequently
due to lack of water supply or loss of water, e.g., by excessive
vomiting and by kidney insufficiency. Low concentration of Na^+ in
plasma is found, for example, by decreased removal of water due to
acute kidney or heart insufficiency, or liver cirrhosis. Low concentration may also be due to intercellular compensation of K^+
deficiency, i.e., diffusion of Na^+ into the cells. A decreased
Na^+ concentration is found in uncomplicated cases with elevated
concentration of glucose (in that way the osmolarity is held constant). The urinary output of Na^+ is less than 10 mmol/day by low
Na^+ intake; by additional steroid insufficiency or alkalosis the
output increases and may appear as "normal."

It is most comprehensive to investigate the intracellular
concentration of *potassium ion,* but in practice the data from plasma
and urine are more easy to obtain and are used in the hospital routine. The K^+ moves out of the cells with a decrease of the intra-

cellular pH and the plasma value will have a tendency to increase. An elevated concentration of K^+ in plasma is found, besides in acidosis, by kidney insufficiency and by cell injuries. Clinical symptoms are found when the concentration is above 7 mmol/liter. Low concentration is found by K^+ loss, e.g., vomiting, diuretica treatment. Clinical symptoms are seen if more than 10% of the whole body K^+ pool is lost. Further information may be found in the literature, e.g., [2].

Lithium ion is a trace element but used orally as a therapeutic for manic-depressive patients [3]. It is easily absorbed and the fate of the component in the organism is nearly that of Na^+. The consequence is that the elimination through the kidney is slow with sodium-poor food and faster with sodium-rich food. The concentration of Li^+ in plasma is estimated as a dosis monitor for the therapy. The therapeutic range is 0.4-1.2 mmol/liter, some side effects may appear above 0.8 mmol/liter. After cessation of such a treatment the halving time of the concentration in serum is normally 1-2 days [3,4].

1.3. Requirement of Analytical Availability

It is essential to follow the water and salt balance for diagnosis and treatment. It must be possible to measure the concentration of Na^+ and K^+ in plasma and urine daily in a clinical chemistry department in the hospital. The determinations might also be carried out on stool in rare cases. Results of Na^+ and K^+ in plasma are so important that they should be available day and night with a short delay time, e.g., 1 hr. Determinations of Na^+ and K^+ are among the most frequent analyses and are carried out on up to several hundred specimens a day. Some criticism of the extensive use of such measurements have, however, been published [5].

Results on Li^+ in plasma are used for dosis monitoring during Li^+ therapy. This analysis is not available in all departments of clinical chemistry but is most frequently carried out in hospitals with a psychiatric department. The analysis is carried out in much smaller series than, say, analysis of Na^+ in plasma, and an immediate result is required in only a few cases.

1.4. Analytical Methods Currently in Use

Chemical methods for the determination of Na^+, K^+, and Li^+ are in general outdated [6]. Flame emission spectrophotometry has been the most popular technique, but also flame absorption spectrophotometry has been used in the routine laboratory. Electrodes are now being used to a greater extent for Na^+ and K^+, partly because this type of measurements requires no special installation or supply in contrast to flame photometry. Flame emission photometry and electrode measurements carried out on dilutions of the plasma or serum give identical results. A direct-electrode measurement in plasma or whole blood will result in higher values due to the plasma water effect, i.e., water replaced by proteins and lipids. Whole-blood measurements have the practical advantage that they can be carried out immediately after the venipuncture, and a centrifugation step is saved. The electrodes estimate the activity of Na^+ and K^+ in the sample corresponding to the concentration of the respective ions in the plasma water of the specimens. A correction for the normal concentration of protein in plasma may be introduced in order to obtain values similar to flame photometry results; the correction is, however, different from one type of equipment to another [9]. The electrode has, however, still the advantage of measuring the biological relevant value in specimens with gross abnormal concentrations of lipids or proteins. It is normal practice in many laboratories to remove the lipids by centrifugation before analysis of strongly lipemic specimen when using flame photometry. A table of corrections of the apparent Na^+ values due to protein and lipids has been published [7].

Neutron activation analysis (see Chap. 6) and combined isotope dilution and mass spectrometry are used for Na^+ and K^+, respectively, as reference methods ("Definitive Methods," [8]).

1.5. Analytical Quality

Preliminary analytical goals for the total analytical variation have been set to $SD/\overline{X} = 0.004$ and $SD/\overline{X} = 0.028$ for Na^+ and K^+ in plasma, respectively. In the case of K^+ these requirements were met in the United States [9] and nearly met in the Scandinavian countries [10]. $SD/\overline{X}$ was, however, found to be 0.015 for Na^+ [10]. Both deviation in accuracy and unsatisfactory precision are the background for that finding. Figure 1 shows the variation in accuracy for 63 Danish laboratories using flame photometry; it is noticed that most labora-tories find slightly smaller values than those estimated by neutron activation. This may be explained by the calibration errors, e.g., the constriction pipets deliver smaller volumes of serum than of water. Furthermore, the basic calibration of the analysis of Na^+ is based on simple aqueous solutions and mistakes may occur. No differences between neutron activation and flame emission spectro-photometry were found when taking this phenomenon into account [11].

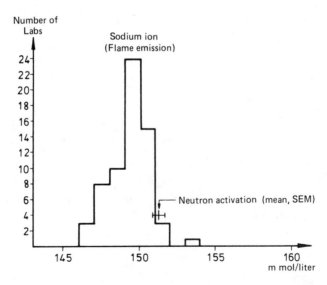

FIG. 1. Variation of accuracy by 63 Danish laboratories. External quality assessment data obtained on lyophilized animal control serum (Autonorm, Nyegaard AS, Norway). Each laboratory measured the mate-rial once a day for 10 days. The average within-laboratory variation was 1.7 mmol/liter.

2. DETERMINATION OF TOTAL CONTENTS OF LITHIUM, SODIUM, AND POTASSIUM SALTS

2.1. Flame Emission Photometry

Flame emission photometry [12-14] is most commonly used for the quantitative measurements of Na^+ and K^+ in body fluids. Li^+, while not present in serum, may also be measured in connection with the therapeutic use of lithium salts in the treatment of some psychiatric disorders.

Atoms of many metallic elements, when supplied with sufficient energy by a hot flame, will emit this energy at wavelengths characteristic for the element. A specific amount or quantum of thermal energy is absorbed by an orbital electron. The electrons, which are unstable in the high energy (excited) state, release their excess energy as photons of a particular wavelength as they change from the excited to their previous or ground state. If the energy is dissipated as light, the light may consist of one or more energy levels and therefore different wavelengths. The line spectra are characteristic for each element. Lithium, sodium, and potassium emit energy primarily at 589, 768, and 671 nm, respectively. The emitted light is isolated (separated from interferring light) by filters, gratings, or prisms.

Alkali metals are comparatively easy to excite in the flame of an ordinary laboratory burner. Thus, flame photometry lends itself well to direct concentration measurements of these metals.

Of the more easily excited alkali metals, such as sodium, only 1-5% of those atoms present in solution become excited in the flame. Even with this small percentage of excited atoms, the method has adequate sensitivity for measurement of alkali metals for most analytical applications.

In some of the earlier instruments for flame emission, standard solutions of sodium or potassium salts were atomized or aspirated directly into the flame. These solutions provided a series of meter readings against which an unknown solution could be compared. This procedure presents several problems:

1. Even small fluctuations in air and gas pressure cause
    unstable response of the instrument and lead to errors.
2. Separate analyses and dilutions must often be made for
    potassium and sodium.
3. The potassium signal is enhanced by the sodium concentra-
    tion in the solution; this effect is known as mutual
    excitation and results from the transfer of energy from
    an excited sodium atom to a potassium ion.

Instead, lithium salt is added to all standards, blanks, and
unknowns in equal concentration, the method being known as the
"internal standard" method. Lithium has a high emission intensity,
is normally absent from biological fluids (cf. page 142), and emits
at 671 nm, which is sufficiently removed from the wavelengths emitted
of sodium and potassium. The flame photometer makes a comparison of
the emission of sodium or potassium with the emission of the reference
element, i.e., lithium. When the ratios of emission are measured in
this way, a compensation for small variations in atomization rates,
flame stability, and solution viscosity is automatically obtained.
Lithium does not function as a "standard" under these conditions but
as a reference element. Lithium also acts as a radiation buffer to
minimize the effects of mutual excitation. The final working concentration of lithium is so high, compared with either sodium or potassium salt, that the same percentage of potassium becomes excited
regardless of the sodium salt concentration in the sample.

2.2. Atomic Absorption Spectrophotometry

Atomic absorption spectrophotometry [14-16] in some respects is the
inverse of flame emission photometry. In all emission methods the
sample is excited in order to measure the radiant energy given off
as the element returns to the lower energy level. Extraneous radiation must be filtered out from the energy of interest if interference
by these signals is to be avoided.

In atomic absorption spectrophotometry the element is not appreciably excited in the flame, but is merely dissociated from its chemical bonds under formation of atoms and thus placed in an unexcited or ground state. This means that the atom is at a low energy level in which it is capable of absorbing radiation at a very narrow band width corresponding to its own line spectrum. A hollow cathode lamp, made of the material to be analyzed, is used to produce light of a wavelength specific for the kind of metal in the cathode. Thus, if the cathode were made of sodium, sodium light at predominantly 589 nm would be emitted by the lamp. When the light from the hollow cathode lamp enters the flame, some of it is absorbed by the ground-state sodium atoms in the flame and in this way exciting some of these atoms. This absorption results in a net decrease in the intensity of the beam from the lamp, and this process is referred to as atomic absorption.

Atomic absorption spectrophotometry is generally considered to be the method of choice for determination of a lithium salt whereas flame emission is used for sodium and potassium salts. Reasonably comparable results for lithium can, however, be obtained by using the method generally used for sodium or potassium, i.e., emission photometry. In this case, potassium salt is used as the internal standard. This method is also subject to interference by other ions as Na^+; therefore, appropriate amounts of this ion must be added to the standard to compensate for this effect.

2.3. Extraction

A highly sensitive and selective extractive spectrophotometric determination of K^+ in the presence of a large excess of sodium has recently been reported in the literature [17]. The principle of this method is based on the selective complex formation of 18-crown-6 (see Fig. 2) with K^+, followed by the solvent extraction of the ion pair formed with bromocresol green. This method has been successfully applied to the determination of potassium salt in serum.

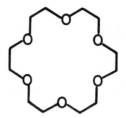

FIG. 2. Structure of 18-crown-6.

3. DETERMINATION OF THE ACTIVITY OF LITHIUM, SODIUM, AND POTASSIUM IONS

3.1. Glass Membrane Electrodes

The various types of ion-selective glass electrodes [18-20] belong
to a continuum of glass electrodes. Changes in the chemical composition of the glass membrane bring about selectivity for cations
other than H^+, e.g., Na^+, K^+, NH_4^+, Ag^+, Cs^+, and Li^+. Alkali metal-
silicate glasses (20% Na_2O/10% CaO/70% SiO_2) containing less than
1% Al_2O_3 produce good pH electrodes with little metal ion response.
Glasses composed of 11% Na_2O/18% Al_2O_3/71% SiO_2 will be highly Na^+-
selective with respect to other alkali ions, e.g., the sensitivity
to Na^+ is much greater than that to K^+. Further reduction in the
concentrations of SiO_2 and Al_2O_3 with a corresponding increase in
Na_2O (68% SiO_2/27% Na_2O/5% Al_2O_3) yields a potassium-to-sodium-ion
selectivity of 20:1. The selectivity of K^+ for the Na^+-selective
electrode mentioned above may be as low as 1:3000 and the electrode
is insensitive to H^+ in the pH range 6-10.

3.2. Liquid Membrane Electrodes

A liquid ion-exchange membrane electrode [20-22] has been shown to
be similar to a glass electrode in that it contains an internal
reference electrode and an internal reference solution of fixed
composition. Instead of glass, the membrane is a thin, porous,

organic polymer saturated with a liquid ion exchanger dissolved in a water-immiscible organic solvent. Valinomycin membranes utilize a neutral carrier and show great K^+ selectivity, about 3800 times that of Na^+ and 18,000 times that of H^+; the electrode can, therefore, be used in strongly acidic media. The valinomycin electrode can be applied for measurement of K^+ in serum.

ABBREVIATIONS

The arithmetic mean value is symbolized by $\overline{X}$, the standard deviation by SD, and the standard error of the mean value by SEM.

REFERENCES

1. H. A. Harper, in *Review of Physiological Chemistry* (H. A. Harper, ed.), Lange, California, 1973, p. 386ff.

2. O. Siggaard-Andersen, in *Chemical Diagnosis of Disease* (S. S. Brown, F. L. Mitchell, and D. S. Yong, eds.), Elsevier/North Holland, 1979, p. 181ff.

3. B. Shopsin and S. Gerskon, *Am. J. Med. Sci., 268,* 306 (1974).

4. A. Amdisen, *Scand. J. Clin. Lab. Invest., 20,* 104 (1967).

5. D. B. Morgan, *Ann. Clin. Biochem., 18,* 275 (1981).

6. N. Weismann and V. J. Pileggi, in *Clinical Chemistry: Principles and Technics* (R. J. Henry, D. C. Cannon, and J. W. Winkelman, eds.), Harper & Row, New York, 1974, p. 81ff.

7. G. B. Levy, *Clin. Chem., 27,* 1435 (1981).

8. N. W. Tietz, *Clin. Chem., 25,* 833 (1979).

9. R. K. Gilbert and R. Platt, *Am. J. Clin. Pathol., 74* (Suppl. 4), 508 (1980).

10. T. Aronson, P. Bjørnstad, E. Leskinen, A. Uldall, and C.-H. de Verdier, in *Assessing Quality Requirements in Clinical Chemistry* (M. Hørder, ed.), The Nordic Clinical Chemistry Project (NORDKEM), Helsinki, 1980, p. 11ff.

11. A. Uldall, in *Nordkem Posters,* The Nordic Clinical Chemistry Project (NORDKEM), Helsinki, 1982.

12. H. H. Willard, L. L. Merritt, Jr., J. A. Dean, and F. A. Settle, Jr., *Instrumental Methods of Analysis,* 6th ed., D. Van Nostrand, New York, 1981, p. 138ff.

13. N. W. Tietz, in *Fundamentals of Clinical Chemistry* (N. W. Tietz, ed.), W. B. Saunders, Philadelphia, 1976, p. 878.

14. W. B. Mason, in *Clinical Chemistry: Principles and Technics* (R. J. Henry, D. C. Cannon, and J. W. Winkelman, eds.), Harper & Row, New York, 1974, p. 51ff.

15. H. H. Willard, L. L. Merritt, Jr., J. A. Dean, and F. A. Settle, Jr., *Instrumental Methods of Analysis,* 6th ed., D. Van Nostrand, New York, 1981, p. 140ff.

16. N. W. Tietz, in *Fundamentals of Clinical Chemistry* (N. W. Tietz, ed.), W. B. Saunders, Philadelphia, 1976, p. 900.

17. H. Sumiyoski, K. Nekahara, and K. Ueno, *Talanta, 24,* 763 (1977).

18. H. H. Willard, L. L. Merritt, Jr., J. A. Dean, and F. A. Settle, Jr., *Instrumental Methods of Analysis,* 6th ed., D. Van Nostrand, New York, 1981, p. 641ff.

19. O. Siggaard-Andersen, in *Fundamentals of Clinical Chemistry* (N. W. Tietz, ed.), W. B. Saunders, Philadelphia, 1976, p. 144.

20. S. M. Reimer, in *Clinical Chemistry: Principles and Technics* (R. J. Henry, D. C. Cannon, and J. W. Winkelman, eds.), Harper & Row, New York, 1974, p. 81ff.

21. H. H. Willard, L. L. Merritt, Jr., J. A. Dean, and F. A. Settle, Jr., *Instrumental Methods of Analysis,* 6th ed., D. Van Nostrand, New York, 1981, p. 645ff.

22. O. Siggaard-Andersen, in *Fundamentals of Clinical Chemistry* (N. W. Tietz, ed.), W. B. Saunders, Philadelphia, 1976, p. 146.

Chapter 8

DETERMINATION OF MAGNESIUM AND CALCIUM IN SERUM

Arne Jensen
Chemistry Department AD
Royal Danish School of Pharmacy
Copenhagen, Denmark

and

Erik Riber
Medi-Lab a.s.
Copenhagen, Denmark

1. CLINICAL CHEMICAL ASPECTS

Calcium is an extremely important element in most living organisms.
Not only does it represent 25% of the bones in vertebrates, but it
has a great influence on secretion processes, on the transmission

of nerve impulses, on the permeability of the cell membrane, and on muscle contractility.

A precise regulation of the activity of Ca^{2+} in serum is therefore of great importance and a complex, integrated system takes care of this regulation.

Three hormones, the parathyroid hormone (PTH), calcitonin, and 1,25-dihydroxy-cholecalciferol $(1,25-(OH)_2D_3)$, are the main hormones in the hormonal regulation of Ca^{2+}. They react via feedback from the concentration of calcium in serum and regulate the calcium concentration via the intestines, the renal tubuli, and the bones.

A schematic representation of the regulatory mechanism is shown in Fig. 1.

The $1,25-(OH)_2D_3$ increases the intestinal absorption of calcium (and magnesium and phosphate).

The PTH increases the synthesis of $1,25-(OH)_2D_3$. It increases the reabsorption of calcium from the tubuli and decreases the phosphate reabsorption. Furthermore, it induces the mobilization of calcium and phosphate from the bones.

Calcitonin works as an antagonist to PTH and $1,25-(OH)_2D_3$.

The hormonal regulation of calcium in serum is controlled by the free calcium ion (Ca^{2+}).

Approximately 45% of the total calcium in serum (100 mg/liter) is bound to the proteins where 80% is bound to albumin; 10% is bound in chelates with citrates, phosphates, etc., and the remaining 45% is Ca^{2+}.

Changes in total serum calcium due to alterations in the serum albumin may consequently occur without changes in the concentration of Ca^{2+}. Changes in the pH of serum will alter the protein-binding capacity. Acidosis will lower the total serum calcium and alkalosis will enhance it.

When evaluating the total serum calcium concentration one must pay attention to the possibility of such "false" deviations from normal.

A possibility is to measure both calcium and albumin and correct the calcium result for deviations in the albumin concentration, or

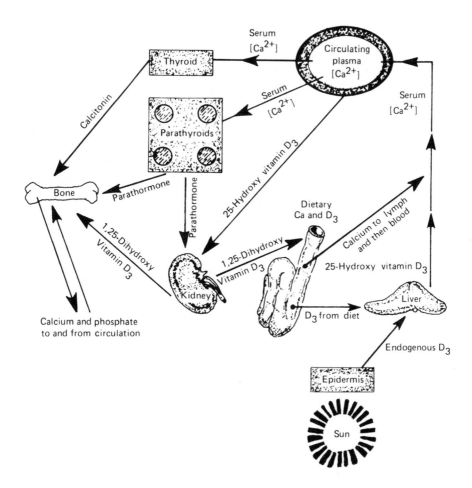

FIG. 1. Schematic representation of those factors maintaining a constant serum calcium level. (Reproduced by permission from Ref. 1.)

better to measure the ultrafiltrable calcium. The optimal parameter is of course the Ca^{2+} concentration and during the last decade this has become more common due to the development of Ca^{2+}-specific electrodes.

Disturbances in the Ca^{2+} balance may occur by diseases in the regulating organs, e.g., the intestines, the kidneys, and the bones. Measurements of PTH, $1,25-(OH)_2D_3$, and calcitonin can in such disorders

be a diagnostic tool, as compensatory changes may occur in the concentration of these hormones.

Disorders in the hormone-producing organs may as well cause
disturbance in the Ca^{2+} balance.

It will be outside the scope of this chapter to go through all
the diseases that involve hyper- and hypocalcemia.

Magnesium has been shown to be the cofactor required for the
utilization of adenosine-5'-triphosphate as a source of energy. It
is thus required for the action of numerous enzyme systems and for
the contraction of muscular tissue.

The adult ingests approximately 300 mg of magnesium daily.
Only 40% of this element is absorbed and excreted in the urine. The
remainder is found in the stool. Serum magnesium levels are found
to be 20 ($\pm$2) mg/liter.

Elevated plasma magnesium level is often observed in oliguric
states, dehydration, and diabetic coma. Magnesium deficiency results
in the drift of calcium out of the bones and in the absence of a pump
mechanism ensues that abnormal calcification takes place in various
organs. Of the magnesium that is filtered through the glomerulous,
95% is generally reabsorbed in the tubule.

Greater amounts of magnesium are retained with decreased
glomerular filtration rate and this results in serum levels above
19-22 mg/liter normally observed.

Magnesium and calcium are, therefore, intimately tied together
in their biological function, and deficiency of either one has a
marked effect on the metabolism of the other. Increasing attention
has thus been paid and is still being paid to losses of magnesium
during fluid therapy and hemodialysis in order to make adequate
replacement of this ion [1].

2. DETERMINATION OF THE TOTAL CONTENTS OF MAGNESIUM AND CALCIUM*

2.1. Atomic Absorption Spectroscopy and Atomic Emission (ICP and Nonflame Methods)

Both magnesium and calcium are determined by means of AAS using an air/acetylene or, in the case of calcium, also a nitrous oxide/ acetylene flame.

Riber [2] determined calcium in serum, and the calibration curve was linear in the range 0-9.0 mg/liter when using a three-slot burner and an air/acetylene flame. This range corresponds to a calcium concentration in serum in the range 0-180 mg/liter after correcting for a dilution factor of 20. The recovery was found to be 99-101%. A matrix containing sulfate and phosphate gave distinctly lower results, but the effect from these two ions could be eliminated completely when 0.5% lanthanum chloride was added. The presence of sodium chloride caused an ionization, but this effect could be removed as well by addition of the $LaCl_3$ solution. The results agree excellently with the results published by Johnson and Reichman [3] and by Osmun [4].

Riber and Persson [5] determined magnesium in serum, and the calibration curve for this metal was linear in the range 0-30 mg/liter when a one-slot burner was used. Recovery experiments were carried out by increasing the concentration of magnesium in serum approximately 20 and 40%, respectively. The recovery was found to be 98-101%. Matrix effects of a similar kind as in the preceding paragraph on Ca were eliminated by addition of La^{3+}.

Berggren [6] used nonflame AAS for the determination of magnesium in pancreatic islets. The method used was injection of aliquots of a homogenized suspension, and the amount found was 0.68 ± 0.15 g/liter.

---

*1 mg/liter = 41 μmol/liter (Mg^{2+}) or 25 μmol/liter (Ca^{2+}).

The homogenized sample was dry-ashed. Effects of various substances on the determination of magnesium have been examined as well. Addition of different sodium salts, potassium chloride, and calcium carbonate gave a relative response of 90-110%.

Addition of $La(NO_3)_3$ gave a relative response of up to 126%, and the presence of $LaCl_3$ might result in a relative response of only 10%.

Speich et al. [7] used AAS for determination of magnesium and calcium in myocardium of 26 control subjects who died of acute trauma and 24 patients who died from acute myocardial infarction. These two elements were estimated using a nitric/hydrochloric Kjeldahl digest. It was found that there was a very significant reversal of the [Mg]/[Ca] ratio in the infarcted area of both groups of patients. Neither detection limit nor sensitivity was stated in connection with this investigation.

Schramel and Klose [8] applied ICP-AES for the determination of magnesium and calcium in human and animal serum (cf. Chap. 9, page 174). A dilution factor of 100 was applied for the determination of magnesium and calcium. The reproducibility was stated to be better than 1% when using ideal experimental conditions. The detection limits for magnesium and calcium were stated to be 10 and 1 μg/liter, and 0.15 and 0.2 μg/liter, the solvents being serum and water, respectively.

Aziz et al. [9] optimized the analytical properties of an injection technique for the analysis of small-volume samples (50-200 μl) by ICP-AES. Samples have been injected into a small funnel connected to a concentric glass nebulizer. When operating the nebulizer at an argon pressure of 5 bar, relative detection limits for calcium and magnesium are a factor of 2-10 higher when compared with ICP methods using continuous nebulization, i.e., Ca(II) showed detection limits 0.2 μg/liter vs. 0.08 μg/liter, and Mg(II) showed similarly 1 μg/liter vs. 0.3 μg/liter. Matrix effects caused by the Na^+ are lower than 10% when sodium salt concentrations are of the order of 5 g/liter. The concentration of magnesium in human serum was by

this technique determined to 22 ± 0.3 mg/liter. Analyses obtained
for a series of control serum samples were for Mg(II), 62 mg/liter
(declared: 74 mg/liter) and for Ca(II), 131 mg/liter (declared:
124 mg/liter). "Cation-cal TM" calibration reference was chosen
as an example.

2.2. Spectrophotometry

Weissman and Pileggi [10] described a method for determination of
magnesium spectrophotometrically in a tungstic acid filtrate of serum
by formation of a red lake with titan yellow in alkaline solution.
Color formation was potentiated by polyvinyl alcohol. The accuracy
of this method has been studied considering the following variables:
calcium, ammonia, iron, phosphate, and pH. It was shown that calcium,
ammonia, and phosphate were without effect up to concentrations of
0.200, 10.000 (as N), and 2.000 (as P) g/liter, respectively, and
that addition of 5 mg of iron(III) per liter to the serum *prior* to
tungstic acid precipitation did not interfere. The amount of NaOH
used in the test could be varied ±10% without effect. Recoveries
of Mg added to serum was 98%, and the precision of the method was
about ±5%.

Tietz [11] described a method for determination of calcium
when o-cresolphthalein was used as (metal) indicator. Serum was
mixed with 0.3 M HCl to dissociate Ca^{2+} from proteins, and Ca^{2+}
(in the sample stream of an autoanalyzer) was then dialyzed into a
reagent stream which contained the metal indicator and 8-hydroxy-
quinoline in dilute HCl. The complex formed was measured at 570 nm.
8-Hydroxyquinoline formed a complex with magnesium ion which might
be present and interference from this ion was therefore prevented
by keeping pH between 10 and 12 by addition of a diethylamine buffer.

Moorehead and Biggs [12] improved the method described for
determination of Ca^{2+} by use of 2-amino-2-methyl-1-propanol at pH
10.0. The authors stated that their method has shown increased
sensitivity, baseline stability, freedom from magnesium interference,

no blanking problems, and finally, it is emphasized that the alcohol
used for the buffer system is neither volatile nor toxic. The results
obtained were not significantly different from AAS results.

Cohen and Sideman [13] proposed a modification of the Moorehead
and Biggs procedure [12], and this modification is claimed to be more
precise and rapid by utilizing small sample size and not requiring
sophisticated laboratory instruments. The modification was validated
by using control serums I and II where the analytical recovery for I
and II was 99.7 and 100.1%, respectively. The results obtained with
the modified technique compared favorably with those obtained with
AAS.

Toffaletti and Kirvan [14] showed that an adapted spectrophotometric method with cresolphthalein complexone has greater detectability
(for dialyzable Ca) than a fluorometric procedure developed by Toffaletti et al. [15]. The coefficient of variation for the spectrophotometric method, as determined from 46 randomized duplicate analyses,
was <1%. The authors concluded that this method was an improvement
with respect to noise level, drift, specificity, detectability, and
more general availability of instrumentation. Moreover, the smaller
sample volume required made possible the routine measurement of
dialyzable calcium in pediatric samples. It must, however, be emphasized that the fluorometric procedure for determination of calcium
is very widely used (cf. Toffaletti et al. [15]).

2.3. Kinetic Analysis

Pausch and Margerum [16] showed that exchange reactions between
Pb(II) and the alkaline earth complexes of CDTA proceeded at sufficiently different velocities to permit the kinetic determination of
mixtures of magnesium and calcium at 260 nm. The determinations were
shown to be rapid and free of interference from other metal ions.
The blood samples were first treated with TCA and filtered. CDTA
was then added and pH adjusted to 9. The general precision for the
procedure was 5-10%. The sensitivity was excellent with determinations

of as little as 0.024 mg/liter of Mg^{2+} and 0.040 mg/liter of Ca^{2+}.
The recovery of the two metals mentioned was 93 and 97%, respectively,
and SD was 1%.

Espersen and Jensen [17] adapted differential kinetic analyses
for magnesium and calcium ions in solution to the flow injection
system [18]. The two-point kinetic assay at 566 and 568 nm, respectively, was based on dissociation of the cryptand (2.2.1) complexes
of magnesium and calcium ions with sodium ion as scavenger. The ions
could be determined with a reproducibility of 5-10% at a sampling
rate of 80 samples/hr with 200-μl samples. The sensitivity was satisfactory with determinations of as little as 5 mg/liter of Mg^{2+} and
8 mg/liter of Ca^{2+}. The recovery was 100-110 and 100-105%, respectively.

Kagenow and Jensen [19] developed a kinetic method for the
simultaneous determination of magnesium and calcium ions in solution
and adapted this method to the stopped-flow injection system [18].
The method was based on the dissociation of the cryptand (2.2.2)
complexes with spectrophotometric monitoring at 575 nm of released
metal ions. Analyses were carried out at 80/hr, and injected sample
volumes were 80 μl. A microcomputer system for data acquisition and
complete control of the stopped-flow injection system was described.
The general precisions for the procedure were 5-10 and 1-5% for the
two metal ions, respectively. The determinations were carried out
with concentrations as low as 1.0 mg/liter of Mg^{2+} and 1.5 mg/liter
of Ca^{2+}. The recovery was 95-110 and 95-105%, respectively.

3. DETERMINATION OF THE ACTIVITY OF MAGNESIUM AND CALCIUM

Fogh-Andersen et al. [20] described a calcium ion-selective electrode
for measurement of free calcium ion in serum at 37°C, based on the
ion exchanger of Ruzicka et al. [21]. The Ca^{2+}-selective membrane
consisted of a solution of calcium di-[di-(n-octylphenyl)phosphate]
in di-(n-octyl)phenylphosphonate, immobilized in a matrix of poly-

vinylchloride. The authors showed that protein interfered on the surface of the unprotected Ca^{2+}-selective membrane and that proteins did not disturb the measurements when the selective membrane was covered by a dialysis membrane.

Siggaard-Andersen et al. [22] recently published a review dealing with the determination of the concentration of free calcium ions in serum. The authors emphasize that with the development of the calcium electrode rapid and precise measurements of the calcium ion activity have become possible, and updated details, as compared to the work by Fogh-Andersen et al. [20], are given regarding these activity measurements.

A plan of the electrochemical cell for measuring the calcium activity is shown in Fig. 2. A mercury-calomel electrode functions as the outer reference electrode and saturated KCl is used as the bridge solution. The sample (X) is via a cellophane membrane in contact with the calcium-sensitive PVC membrane. The inner solution consists of an aqueous calcium chloride with a constant Ca^{2+} activity, and the cell is completed by the inner Ag/AgCl reference electrode.

The electrode used by Siggaard-Andersen et al. [22] is also based on the ion exchanger of Ruzicka et al. [21]. The electrode covers a measuring range for the concentration of Ca^{2+} from 4.0 mg/liter to 40 g/liter, i.e., a Nernstian response is found in this

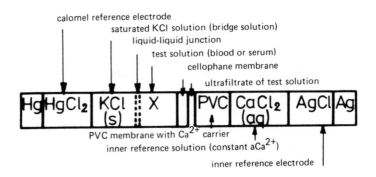

FIG. 2. Schematic representation of the cell with the calcium electrode. (Reproduced by permission from Ref. 22.)

range, or the electrode sensitivity closely approaches the theoretical one (30.77 mV per decade at 37°C).

The selectivity of the electrode is so good that ions in serum (apart from Ca^{2+}) do not cause a bias of more than at most 0.5% even if the ions are present in pathological concentrations. It is emphasized that the membrane can be damaged by organic solvents even in low concentrations and that the electrode, therefore, cannot be used in these solvents.

It is stressed in this chapter that protein-containing solutions cause a minimal drift of 0-0.2 mV and alternating measurements in protein-free and protein-containing solutions become possible *if* the calcium ion-sensitive PVC membrane is covered by a cellophane membrane. At the same time, however, the response time of the electrode increases from less than 1 sec to about 15 sec for 95% response.

When measurement of the calcium activity is performed on serum, the blood is collected in a dry test tube, which is then stoppered, with as little air as possible, in order to limit the loss of CO_2 to a minimum. The tube is left for a half to one hour and serum is subsequently separated by centrifugation. If a loss of CO_2 has occurred so that the pH of the serum is higher than 7.6, the serum should be reequilibrated with a CO_2-air mixture in order to bring the pH within the interval 7.2-7.6. The concentration (or activity) of Ca^{2+} and pH are then measured, and the Ca^{2+} value is then corrected to pH = 7.40 according to standard procedures.

Siggaard-Andersen et al. [22] state that the precision of measurements with the calcium electrode, calculated from duplicate measurements on a series of samples, give a coefficient of variation of 1.1% for the concentration (or activity) of Ca^{2+} measured in serum and referred to pH = 7.40. It is furthermore concluded that the accuracy of measurements is high, mainly on the basis of the Nernstian response of the electrode and the excellent selectivity, but the accuracy as such is difficult to evaluate because no reference method is available.

A summary of the methods used for the determination of magnesium and calcium is given in Table 1.

TABLE 1

Methods Used for the Determination of Magnesium and Calcium

Element	Method	Detection limit (μg/liter)	Sensitivity (mg/liter)	Standard deviation	Pretreatment of sample analytical range	Biological material	Ref.
Ca	AAS			0.7 mg/liter	Dilution with water, 20x 0-180 mg/liter	Serum	[2]
Mg	AAS			0.3 mg/liter	Direct determination 0-30 mg/liter	Serum	[5]
Mg	Flameless AAS			150 mg/kg	Homogenized suspension	Pancreatic islets	[6]
Mg } Ca	AAS				Wet digestion	Myocardium	[7]
Mg } Ca	ICP-AES	10 / 1			Dilution with water, 100x	Serum	[8]
Mg } Ca	ICP-AES	{x1 *0.3} {x0.2 *0.08}			Dilution with Herrmann solution, 1+4 — 18-62 mg/liter (Mg) — 78-131 mg/liter (Ca)	Serum	[9]

(x Injec. of sample)
(* Cont. nebulizat.)

Analyte	Method	Concentration	Sample preparation	Sample	Ref.
Mg	Spectrophotometry		Tungstic acid filtrate 0-32 mg/liter	Serum	[10]
Ca	Spectrophotometry		Acidification with 0.3 M HCl	Serum	[11]
Ca	Spectrophotometry		Acidification with HCl	Serum	[12]
Ca	Spectrophotometry	1.1-1.7 mg/liter	Acidification with HCl	Serum	[13]
	Spectrophotometry	5.6 mg/liter	Addition of buffer and dialysis 51-57 mg/liter	Serum	[14]
Mg / Ca	Kinetic (spectrophotometry)	0.024 0.040	Treatment with TCA and filtration	Serum	[16]
Mg / Ca	Kinetic (spectrophotometry)	5 8		Aqueous solution	[17]
Mg / Ca	Kinetic (spectrophotometry)	1.0 1.5		Aqueous solution	[19]
Ca	Potentiometry (activity)		4.0 mg/liter-40 g/liter	Serum	[22]

163

ABBREVIATIONS AND DEFINITIONS

AAS	atomic absorption spectroscopy
ADP	adenosine-5'-diphosphate
AES	atomic emission spectroscopy
ATP	adenosine-5'-triphosphate
CDTA	trans-1,2-diaminocyclohexane--N,N,N',N'-tetraacetate
Detection limit	concentration that gives a signal-to-noise ratio of 2
EDTA	ethylenediamine tetraacetate
ICP	inductively coupled plasma
PTH	parathyroid hormone
PVC	polyvinylchloride
SD	standard deviation
Sensitivity	concentration that produces an absorbance of 0.0044 (1% absorption)
TCA	trichloroacetic acid

ACKNOWLEDGMENT

The authors thank Dr. Finn Christensen, M.D., Ph.D., for his critical review of the manuscript.

REFERENCES

1. S. Natelson and E. Natelson, in *Principles of Applied Clinical Chemistry*, Vol. 1, *Maintenance of Fluid and Electrolyte Balance*, Plenum Press, New York, 1975, p. 141.

2. E. Riber, unpublished results.

3. J. R. K. Johnson and G. C. Reichman, *Clin. Chem.*, *14*, 1218 (1968).

4. J. R. Osmun, *Am. J. Med. Tech.*, *33*, 448 (1967).

5. E. Riber and P. S. Persson, unpublished results.

6. P.-O. Berggren, *Anal. Chim. Acta, 119,* 161 (1980).

7. M. Speich, B. Bousquet, and G. Nicolas, *Clin. Chem., 26*(12), 1662 (1980).

8. P. Schramel and B.-J. Klose, *Z. Anal. Chem., 307,* 26 (1981).

9. A. Aziz, J. A. C. Broekaert, and F. Leis, *Spectrochim. Acta, 36B,* 251 (1981).

10. N. Weissman and V. J. Pileggi, in *Clinical Chemistry: Principles and Technics, Inorganic Ions* (R. J. Henry, D. C. Cannon, and J. W. Winkelman, eds.), Harper & Row, New York, 1974, 2nd ed., p. 673.

11. N. W. Tietz, in *Fundamentals of Clinical Chemistry, Electrolytes* (N. W. Tietz, ed.), W. B. Saunders, Philadelphia, 1976, p. 908.

12. W. R. Moorehead and H. G. Biggs, *Clin. Chem., 20*(11), 1458 (1974).

13. S. A. Cohen and L. Sideman, *Clin. Chem., 25*(8), 1519 (1979).

14. J. Toffaletti and K. Kirvan, *Clin. Chem., 26*(11), 1562 (1980).

15. J. Toffaletti, J. Savory, and H. J. Gitelman, *Clin. Chem., 23,* 1258 (1977). Cit. from Ref. 14.

16. J. B. Pausch and D. W. Margerum, *Anal. Chem., 41,* 226 (1969).

17. D. Espersen and A. Jensen, *Anal. Chim. Acta, 108,* 241 (1979).

18. J. Ruzicka and E. H. Hansen, *Flow Injection Analysis,* Wiley-Interscience, New York, 1981.

19. H. Kagenow and A. Jensen, *Anal. Chim. Acta, 145,* 125 (1983).

20. N. Fogh-Andersen, T. F. Christiansen, L. Komarmy, and O. Siggaard-Andersen, *Clin. Chem., 24*(9), 1545 (1978).

21. J. Ruzicka, E. H. Hansen, and J. C. Tjell, *Anal. Chim. Acta, 67,* 155 (1973).

22. O. Siggaard-Andersen, J. Thode, and J. Wandrup, in *Blood pH, Carbon Dioxide, Oxygen and Calcium-Ion, The Concentration of Free Calcium Ions in the Blood Plasma "Ionized Calcium"* (O. Siggaard-Andersen, ed.), Private Press, Copenhagen, 1981, p. 163.

Chapter 9

DETERMINATION OF MANGANESE, IRON, COBALT, NICKEL,
COPPER, AND ZINC IN CLINICAL CHEMISTRY

Arne Jensen
Chemistry Department AD
Royal Danish School of Pharmacy
Copenhagen, Denmark

Erik Riber and Poul Persson
Medi-Lab a.s.
Copenhagen, Denmark

and

Kaj Heydorn
Isotope Division
Risø National Laboratory
Roskilde, Denmark

1. CLINICAL CHEMICAL ASPECTS

Absorption of manganese in humans has been shown to be only a few
percent, but it has been observed that an increased absorption takes
place with iron deficiency. Highest concentrations were found in the
liver and kidney, but the concentration in the brain was low. The
gastrointestinal tract has been shown to be the major route of elimination. Biological half-time in humans was shown to be 2-5 weeks,
depending on body stores. Manganese has been shown to be an essential element, but manganese deficiency has not generally been found.
Excessive exposure via inhalation has been shown to cause damage to
the lungs. Excessive absorption in the lungs has caused accumulation
in the brain. Manganese has often been seen to cause an irreversible
brain disease, to some extent similar to Parkinson's disease. However, manganese concentrations decreased after cessation of exposure
[1].

It is well known that iron is essential and takes part in
oxygen transport and utilization. Iron deficiency has been observed
very commonly, especially among premenopausal women. Only 0.01%/day
of absorbed iron is eliminated. Ingestion of soluble iron salts by
children in doses exceeding 0.5 g of iron has given rise to severe
lesions in the gastrointestinal tract and, later, metabolic acidosis,
shock, and toxic hepatitis. Long-term ingestion of excessive amounts
of iron has caused hemosiderosis and, in severe cases, cirrhosis of
the liver. Occupational exposure to iron oxides during mining or
refining of iron ore sometimes has caused hemosiderosis of the lung.
Increased mortality in lung cancer has repeatedly been reported from
hematite mines and iron foundries. It is suspected that iron oxide
dust might serve as a cocarcinogenic substance, i.e., enhancing the
development of cancer at a simultaneous exposure to carcinogenic
substances [2].

Cobalt is essential as an integral component of vitamin B_{12}.
Gastrointestinal absorption of cobalt has been estimated to be about
25%, with wide individual variation, and the excretion has been shown
to take place mainly via the urinary tract. It has been established

that a major part of cobalt is excreted within days and the rest of Co with a biological half-time of at least a couple of years. Addition of cobalt to beer has caused endemic outbreaks of cardiomyopathy among heavy beer drinkers, resulting in a number of fatalities. Myocardial degeneration and electrocardiographic changes have also been seen in laboratory animals after repeated parenteral or peroral exposure to cobalt. Industrial exposure has given rise to pneumoconiosis and other pulmonary effects [3].

Nickel has so far not been proven to be essential for humans, but it has been recognized as an essential trace element for higher animals. Inorganic nickel compounds are absorbed to a few percent from the gastrointestinal tract. Absorption from the lungs has been shown to depend on solubility. Nickel is accumulated in kidneys, liver, and lungs. The excretion has been established as rapid and as taking place mainly via the urine. Nickel carbonyl has high toxicity, and the lungs are the critical organ after both inhalation and injection. This nickel complex is then decomposed to nickel and carbon monoxide, and after oxidation Ni is excreted via the kidneys. Dermatitis due to nickel has been observed to be common among nickel workers as well as in the general population. Exposure to nickel carbonyl has given rise to a large number of acute poisonings, often lethal, with pulmonary damage as the main feature. Occupational exposure to nickel compounds with low solubility has caused cancer of the lungs and, in some cases, also of the nasal tissue [4].

Copper is an essential metal. The absorption of ingested copper in food has been established to be about 50% and normally to be regulated by homeostatic mechanisms. The storage of copper has been found to take place in liver and muscles. Excretion of this element has been observed mainly via the bile and only a few percent of the absorbed amount has been found in the urine. The biological half-time in humans has been estimated at about 4 weeks. Ingestion of larger amounts of copper salts by accident has nearly always caused gastrointestinal disturbances including vomiting. Inhalation of copper fumes has often caused metal fume fever. Chronic copper

effects have never been reported, even if copper exposure is common
in industry [5].

Zinc is also an essential metal and indispensable for the
function of various enzymes. Zinc deficiency has been observed in
human beings. High zinc concentrations have been found in prostate,
bone, muscle, and liver, and the excretion occurs primarily via the
gastrointestinal tract. The biological half-time of retained zinc
has been proven to be about 1 year. Large oral doses of zinc salts
have caused gastrointestinal disorders including vomiting and diarrhea. The emetic dose of zinc salt has been estimated to be about
300 mg. Metal fume fever has been observed as an acute disorder
after respiratory exposure to freshly generated zinc fumes (mostly
as ZnO). Exposure to high concentrations of atmospheric $ZnCl_2$ has
often been fatal and has included acute damage to the mucous membranes of nasopharynx and respiratory tract [6].

2.   DETERMINATION OF THE TOTAL CONTENTS OF MANGANESE, IRON, COBALT, NICKEL, COPPER, AND ZINC*

2.1.   Spectrophotometry

Determination of the metals treated in this chapter is in general not
carried out spectrophotometrically. There are, however, a few examples. About 30 years ago, Gubler et al. [7] described a method for
determination of copper in whole blood, red cells, and plasma. The
copper was liberated from the proteins with HCl and TCA. The copper
in the supernatant was then determined spectrophotometrically by means
of sodium diethyldithiocarbamate. The recovery was 98-101% of the
added copper and the detection limit could be estimated to 100 μg/
liter.

Giovanniello and Pecci [8] described both the classical determination of serum iron and total iron-binding capacity (TIBC) by

---

*1 mg/liter = 0.0182 mmol/liter (Mn), 0.0179 mmol/liter (Fe), 0.0170
mmol/liter (Co), 0.0170 mmol/liter (Ni), 0.0157 mmol/liter (Cu), and
0.0153 mmol/liter (Zn).

means of sulfonated bathophenanthroline as an automatic procedure for
this analysis. Gantcheva and Bontchev [9] described the catalytic
oxidation of THB with hydrogen peroxide in the presence of Cu(II) and
have applied the results obtained for development of a catalytic procedure for determination of the copper content in blood serum. The
oxidation of THB was followed spectrophotometrically (480 nm). The
kinetic method had no systematic errors, and a comparison of this
method with photometric and AAS methods showed that the reproducibility and accuracy of the kinetic method are of the same order. The
same authors have in a supplementary publication elucidated the mechanism of the catalyzed oxidation of THB.

Mori et al. [10] modified the sensitive method for serum iron
and total iron-binding capacity in which the color reagent ferrozine
is used. The method has been modified and adapted to different discrete analyzers and in Ref. 10 to the Abbott discrete analyzer. The
standard curve was shown to be linear to at least 10 mg/liter.
Analytical recovery was 97-100%.

2.2. Voltammetric Methods

The ions of the elements under discussion can all be determined by
means of voltammetric methods; however, the methods have only been
used to a limited degree in clinical chemistry as far as the metals
treated in this chapter are concerned (cf. also Chap. 10). Franke
and Zeeuw [11] determined Zn and Cu, among many other metals, in
urine without prior destruction, thus providing a tremendous gain in
analysis time. Blood analysis could also be carried out, but it was
shown that blood needed destruction before the ASV. The screening
was shown to be very sensitive, allowing determination down to the
lower parts per billion range.

Pinchin and Newham [12] investigated the determination of Cu,
among other metals, by ASV at a wax-impregnated graphite electrode,
preplated with mercury. Calibration graphs were not linear for low
concentrations of Cu. Whole-blood samples were treated with $HClO_4$,

digested for 1.5-2 hr, and finally adjusted to a volume of 5 ml at
pH 5. The recovery was found to be 99% at a concentration of 10
µg/liter.

Flora and Nieboer [13] determined Ni by DPP at a DME. DMG
sensitized the DPP method. The detection limit was 2 µg/liter and
the linear response was in the concentration range 0-85 µg/liter,
but this could easily be extended by omitting DMG. Actual analyses
of only plant tissue, and not biological material as such, have been
reported.

Wang [14] evaluated ASV at microelectrodes in vitro toward its
exploitation for in vivo measurements of some trace metals, e.g.,
copper. Well-defined stripping voltammograms were obtained for this
metal at the 50-80 µg/liter level using short (1-5 min) deposition
periods. The stripping at a bare carbon surface was compared with
that of a preplated mercury film. The effect of dissolved oxygen
on the stripping voltammograms was evaluated, and it was shown that
improved sensitivity and detectability in the presence of oxygen
might be obtained by adopting stripping modes such as subtractive
ASV [15] or PSA [16], which have been shown to be much less affected
by dissolved oxygen. The author concluded by emphasizing that major
experimental difficulties still remained before it would be possible
to apply ASV for analysis of trace metals in vivo and that different
effects were, in the hour of writing, being examined, e.g., complexation and adsorption of various proteins on ASV peak currents and the
corresponding background current.

2.3. Atomic Absorption Spectroscopy and Atomic Emission
   (Classical, Nonflame, and ICP Methods)

This section will, in chronological order, cover a representative
part of the literature, dealing with the determination in clinical
chemistry of the elements, mentioned in the title of this chapter,
by AAS and AES-ICP methods. Only the last 2-3 years will be covered,
due to the large number of papers published.

Lindh et al. [17] determined copper and zinc, among several other elements, in bone tissue from autopsy specimens of the femur of workers who had been exposed to a large number of metals in a smeltery and refinery. In addition, a control group was quantitatively assayed. The analytical techniques used were flameless AAS, NAA, and particle-induced x-ray emission analysis in a proton microprobe. Levels of copper and zinc were found to be 2-8 and 71-137 mg/kg, respectively, i.e., the medium level of these two elements in the group of exposed workers did not exceed the corresponding value of the control group.

Nakamura et al. [18] described the determination of iron in serum by flameless AAS using a graphite furnace atomizer. The serum was diluted 40 times with water and injected into the graphite tube. The procedure required no sample pretreatment. The method described could determine any species of iron in serum. The level of iron in human serum found by this method was 0.76-1.72 mg/liter, and the recovery of the present method was 82-111%.

Taylor and Bryant [19] compared methods for preparation of serum samples and for determination of copper and zinc by AAS by using the results obtained from an external quality control scheme. The methods most commonly employed were dilution with water, dilution with butan-1-ol or propan-1-ol, removal of protein by TCA precipitation, followed by flameless AAS. The effects of the inclusion of sodium and potassium ion on the results have been examined. Low results for serum copper were found by flameless AAS. High zinc values were found following TCA precipitation. None of the methods showed superior quality for between-batch precision. The ionization interference was variable depending on the method of dilution, and the addition of sodium and potassium to standard solutions was not always necessary. In conclusion: the results from the quality control scheme indicated that the measurement of serum zinc was technically difficult compared with serum copper, and this was stated to be probably due to contamination. The wide variance of zinc results obtained for each sample, therefore, hindered the identification of

any methodological features which had a significant effect on the
analysis of this element.

Bagliano et al. [20] determined several trace metals in hair
by AAS, among these zinc. Acid digestion procedures are described
for the dissolution of human hair in routine determinations, and
zinc is determined by flame AAS using air/acetylene. For hair, the
concentration range, limit of detection, and sensitivity were 50-300,
15.0, and 26 mg/kg, respectively.

Vieira and Hansen [21] described a precise flameless AAS method
requiring only 10 µl of sample (serum and urine) which thus has given
the opportunity for repeated measurements in the neonate. Standard
curves covered the range of 0-30 mg/liter. Environmental zinc contamination was controlled by using appropriate disposable plastic
ware, thoroughly deionized water, etc. Matrix effects for urine
determinations were variable, which required that the method of
standard additions be used. The matrix effects for serum and plasma
were compensated for by adding to the standard albumin, serum, or
plasma, each of which had been dialyzed to low zinc concentrations.
The recovery was 101-102%. Detection limit and sensitivity were not
included in this paper.

Pleban et al. [22] used polarized Zeeman effect flameless AAS
for the quantitative measurement of manganese and copper of human
kidney cortex (cf. Chap. 10, Ref. 25). Within-run coefficients of
variation for manganese and copper, 56.3 and 177.2 µg/liter, respectively, were 5.6 and 6.3%, respectively. For manganese and copper
were found digest concentrations (wet digestion, mean ± SD) of
20.9 ± 0.4 and 41.4 ± 2.6 µg/liter, respectively, and detection
limits were 5.2 µg/liter for copper and 0.8 µg/liter for manganese.
Average recoveries for the two metals ranged between 95 and 101%.
The standard curves were shown to be linear to 200 µg/liter for
copper and to 100 µg/liter for manganese.

Schramel and Klose [23] applied ICP-AES for the determination
of Fe, Cu, and Zn. Chemical interferences did not appear in this
technique due to the very high excitation temperature. The concentrations of the elements in pig serum were found to be 2.3 ± 0.5

mg/liter, 2.0 ± 0.2 mg/liter, and 1.0 ± 0.2 mg/liter for Fe, Cu, and Zn, respectively (average and SD, n = 30). The sensitivity of the determinations of Fe, Cu, and Zn in serum were 10, 10, and 5 µg/liter, respectively.

Blacklock and Sadler [24] developed a semiquantitative screening method for heavy metals (including Fe, Co, Ni, Cu, and Zn) using a rapid screening method by emission spectroscopy (ES) (cf. Chap. 10, Ref. 29).

Gardiner et al. [25] used gel filtration with Sephadex G-100 to partition zinc- and copper-containing proteins in samples of human sera. The zinc and copper content in the fractions collected have been determined by flameless AAS without sample pretreatment. The results have confirmed the known association of zinc with $\alpha2$-macroglobulin and albumin, but association with other proteins has also been found.

Goldberg and Allen [26] measured copper, manganese, and iron in wet-solubilized brain tissue. Copper and manganese concentrations were determined by using flameless AAS and iron concentration by AAS (air/acetylene flame). The basal ganglia contained high concentrations of iron and copper as compared with concentrations found in the cortex or the hippocampus. Manganese was uniformly distributed in the regions examined.

Halls et al. [27] described two methods for the determination of copper in urine by flameless AAS. The first, suitable for spectrometers capable of making good background correction, involved a direct determination against simple aqueous standards after 2x dilution of the sample. The second required no background correction and could be used when background correction was not available or was inadequate. The copper was extracted from the urine with APDC into MIBK and the extract was analyzed for copper. No significant matrix effect was observed for either method. The pretreatment was dry ashing. The recovery was 93-106 and 89-96% for the direct and extraction methods, respectively. Detection limits and sensitivities had not been determined, according to the authors. The precision was shown

TABLE 1

Methods Used for the Determination of Manganese, Iron, Cobalt, Nickel, Copper, and Zinc

Element	Method	Detection limit	Sensitivity	Standard deviation	Pretreatment of sample and analytical range	Biological material	Ref.
Cu	Spectrophotom.	100 µg/liter			Liberation of Cu with HCl and TCA 450-17,200 µg/liter	Whole blood	[7]
Fe	Spectrophotom.				Liberation of Fe with HCl and TCA (serum iron)	Serum	[8]
					Addition of excess Fe and removal of excess or unbound Fe with anion-exchange resin (TIBC)		
					0.56-1.83 mg/liter (serum iron) 2.48-4.22 mg/liter (TIBC)	Serum	
Cu	Spectrophotom.	1 µg/liter			Liberation of Cu with HCl and TCA	Serum	[9]
Fe	Spectrophotom.				Cf. Ref. 8. 1.17-1.91 mg/liter (serum iron) 2.67-3.84 mg/liter (TIBC)	Serum	[10]
Cu, Zn	ASV	1-3 µg/liter			Direct determ.	Urine	[11]
Cu	ASV				Wet ashing	Whole blood	[12]
Ni	DPP	2 µg/liter			Dry ashing	Plant tissue	[13]
Cu	ASV	300 ng/liter			Aqueous solutions	Aqueous solutions	[14]
Cu, Zn	Flameless AAS				Wet ashing 2-8 mg/kg (Cu) 71-137 mg/kg (Zn)	Bone tissue	[17]

Element	Method				Sample preparation	Sample	Ref.
Fe	Flameless AAS				Dilution of serum with water (40x) 0.76-1.72 mg/liter	Serum	[18]
Cu, Zn	Flameless AAS				Dilution or TCA precipitation	Serum	[19]
Zn	Flameless AAS	15 mg/kg	26 mg/kg		Wet digestion 50-300 mg/kg	Hair	[20]
Zn	Flameless AAS				Dilution with water (100x) 0-30 mg/liter	Serum, urine	[21]
Mn, Cu	Flameless AAS + pol. Zeeman eff.	0.8 µg/liter 5.2 µg/liter	0.4 µg/liter 2.6 µg/liter		Wet digestion 0-100 µg/liter 0-200 µg/liter	Human kidney cortex	[22]
Fe Cu Zn	ICP-AES		10 µg/liter 10 µg/liter 5 µg/liter	0.5 mg/liter 0.2 mg/liter 0.2 mg/liter	Direct determ.	Pig serum	[23]
Fe, Co Ni, Cu, Zn	ES				Wet digestion of tissue, extraction	Urine, blood, tissue from rats	[24]
Zn, Cu	Flameless AAS	0.31 µg/liter 1.29 µg/liter	0.47 µg/liter 1.08 µg/liter	0.15 µg/liter 0.65 µg/liter	Direct determ. 0-40 µg/liter 0-80 µg/liter	Serum	[25]
Cu, Mn Fe	Flameless AAS AAS (air, C_2H_2)				Wet digestion	Brain tissue	[26]
Cu	Flameless AAS				1) Dry ashing Direct determ. (dilution of urine 2x) 2) Extraction 4-67 µg/liter	Urine	[27]
	NAA				Cf. Tables 2 and 3		

to be better for the direct method as there was less manipulation of the sample.

The different methods which have been described so far in this chapter and which were used for the determination of manganese, iron, cobalt, nickel, copper, and zinc are summarized in Table 1.

2.4. Neutron Activation Analysis

All six elements under discussion in this chapter may in principle be determined by thermal neutron activation analysis, and the respective indicators are listed in Table 2 together with their interference-free detection limits [28]. The analytical methodology is described in Chap. 6, in which the factors affecting precision, accuracy, and sensitivity are discussed. In practice the detection limit for nickel is too high to permit its determination in normal biomedical samples.

TABLE 2

Indicators for the Determination of Some Essential Trace
Elements by Thermal Neutron Activation Analysis

Element	Nuclide	Half-life	γ energy (keV)	Detection limit (ng)[*]
Manganese	^{56}Mn	2.6 hr	847	0.001
Iron	^{59}Fe	45.6 days	1099	3200
Cobalt	^{60}Co	5.3 years	1173	12
Nickel	^{65}Ni	2.6 hr	367	7
Copper	^{64}Cu	12.8 hr	511	0.035
Zinc	^{65}Zn	245 days	1115	420

[*]After 5 hr irradiation at a thermal neutron flux density of 10^{17} n/(m$^2 \cdot$s) according to Guinn and Hoste [28].

2.4.1. *Radiochemical Separation*

The detection limits quoted in Table 2 for iron, cobalt, and zinc
are also high in relation to normal samples but may be reduced so
much by increasing the irradiation time that these elements, together
with copper and manganese, can all be determined at their normal concentrations in biomedical samples after radiochemical separation.

These methods require decomposition of the sample after irradiation with thermal neutrons, and the addition of carrier not only
facilitates separation but also permits the determination of chemical
yield. Simultaneous determination of several elements in the same
sample, such as manganese, copper, and zinc, has been performed at
the University of Gent [29] and at Risø National Laboratory [30].
These methods are applicable to the analysis of tissue samples, as
well as blood cells and plasma, where manganese in particular has
given rise to considerable difficulties [31].

2.4.2. *Instrumental Neutron Activation Analysis*

Elements with long-lived indicators, in particular zinc and cobalt,
may be determined in biomedical samples without any radiochemical
separation by leaving the irradiated sample for several months, until
all other radionuclides have decayed away. This nondestructive or
instrumental method may also be applied to the determination of manganese and iron, but only in samples with comparatively high concentrations. This approach has been adopted by several groups in Sweden
[32], Germany [33], and Japan [34].

Results for copper by instrumental neutron activation analysis
are less reliable because the γ energy characterizing the ^{64}Cu indicator is also produced by several other radionuclides.

2.4.3. *Examples*

Recent examples of the use of neutron activation analysis in biomedical research are listed in Table 3. Additional examples may be

TABLE 3

Some Recent Examples of the Use of Neutron Activation
Analysis for the Determination of Five Essential
Trace Elements in Biomedical Samples

Tissue	Manganese	Iron	Cobalt	Copper	Zinc
Blood	[29,37]	[33,34]	[33,34,46]	[29,34]	[29,33,34]
Urine	[38]		[38,43]	[38]	[38,43]
Brain	[39,42]	[42]		[44]	[42]
Heart	[39,42]	[32,42]	[32]	[32]	[32,40,42]
Kidney ⎱ Liver ⎰	[32,40,41, 42]	[32,40,42]	[32,40]	[32,40,44]	[32,40,42, 45]
Lung	[40,42]	[40,42]	[40]	[40,44]	[40,42]
Muscle	[41,42]	[42]		[44]	[42]

found in the review by Bowen [35], and with special reference to
copper and zinc in an article by Carden and Fink [36].

ABBREVIATIONS AND DEFINITIONS

AAS	atomic absorption spectroscopy
AES	atomic emission spectroscopy
APDC	ammonium pyrrolidone dithiocarbamate
ASV	anodic stripping voltammetry
Detection limit	concentration that gives a signal-to-noise ratio of 2
DME	dropping mercury electrode
DMG	dimethyl glyoxime
DPP	differential pulse polarography
ES	emission spectroscopy (using a grating spectrograph)
ICP	inductively-coupled plasma
MIBK	methyl isobutyl ketone

NAA	neutron activation analysis
PSA	potentiometric stripping analysis
Sensitivity	concentration that produces an absorbance of 0.0044 (1% absorption)
SD	standard deviation
TCA	trichloroacetic acid
THB	1,3,5-trihydroxybenzene
TIBC	total iron-binding capacity

REFERENCES

1. M. Piscator, in *Handbook on the Toxicology of Metals* (L. Friberg, G. F. Nordberg, and V. B. Vouk, eds.), Elsevier/North Holland, Amsterdam, 1979, p. 485.

2. C.-G. Elinder and M. Piscator, in Ref. 1, p. 435.

3. C.-G. Elinder and L. Friberg, in Ref. 1, p. 399.

4. T. Norseth and M. Piscator, in Ref. 1, p. 541. See also F. H. Nielsen, Essentiality and function of nickel, in *Trace Element Metabolism in Animals* (W. G. Hoekstra, J. W. Suttie, H. E. Ganther, and W. Merty, eds.), Vol. 2, Univ. Park Press, Baltimore, 1974, pp. 381-385.

5. M. Piscator, in Ref. 1, p. 411.

6. C.-G. Elinder and M. Piscator, in Ref. 1, p. 675.

7. C. J. Gubler, M. J. Lahey, H. Ashenbrucker, G. E. Cartwright, and M. M. Wintrobe, *J. Biol. Chem., 196,* 209 (1952).

8. T. J. Giovanniello and J. Pecci, in *Standard Methods of Clinical Chemistry* (G. R. Cooper and J. Stanton King, Jr., eds.), Academic Press, New York, Vol. 7, 1972, p. 127.

9. S. Gantcheva and P. R. Bontchev, *Talanta, 27,* 893 (1980).

10. L. Mori, A. Bekkering, J. Traini, and L. Vanderbinden, *Clin. Chem., 27*(8), 1441 (1981).

11. J. P. Franke and R. A. de Zeeuw, in *Clinical Chemistry and Chemical Toxicology of Metals* (S. S. Brown, ed.), Elsevier/North Holland, Amsterdam, 1977, p. 371.

12. M. J. Pinchin and J. Newham, *Anal. Chim. Acta, 90,* 91 (1977).

13. C. J. Flora and E. Nieboer, *Anal. Chem., 52,* 1013 (1980).

14. J. Wang, *Anal. Chem., 54,* 221 (1982).

15. J. Wang and M. Ariel, *Anal. Chim. Acta, 128,* 147 (1981).

16. D. Jagner, M. Josefson, J. Westerlaun, and K. Åren, *Anal. Chem.*, *53*, 1406 (1981). Cit. from Ref. 14.

17. U. Lindh, D. Brune, G. F. Nordberg, and P.-O. Wester, *Sci. Total Environ.*, *16*, 109 (1980).

18. K. Nakamura, H. Watanabe, and H. Orii, *Anal. Chim. Acta*, *120*, 155 (1980).

19. A. Taylor and T. N. Bryant, *Clin. Chim. Acta*, *110*, 83 (1981).

20. G. Bagliano, F. Benischek, and I. Huber, *Anal. Chim. Acta*, *123*, 45 (1981).

21. N. E. Vieira and J. W. Hansen, *Clin. Chem.*, *27*(1), 73 (1981).

22. P. A. Pleban, J. Kerkay, and K. H. Pearson, *Clin. Chem.*, *27*(1), 68 (1981).

23. P. Schramel and B.-J. Klose, *Fresenius' Z. Anal. Chem.*, *307*, 347 (1981).

24. E. C. Blacklock and P. A. Sadler, *Clin. Chim. Acta*, *113*, 87 (1981).

25. P. E. Gardiner, J. M. Ottoway, G. S. Fell, and R. R. Burns, *Anal. Chim. Acta*, *124*, 281 (1981).

26. W. J. Goldberg and N. Allen, *Clin. Chem.*, *27*(4), 562 (1981).

27. D. J. Halls, G. S. Fell, and P. M. Dunbar, *Clin. Chim. Acta*, *114*, 21 (1981).

28. V. P. Guinn and J. Hoste, in *Elemental Analysis of Biological Materials* (Technical Report Series No. 197), International Atomic Energy Agency, Vienna, 1980, p. 105ff.

29. J. Versieck, J. Hoste, and F. Barbier, *Acta Gastroenterol. Belg.*, *39*, 340 (1976).

30. E. Damsgaard and K. Heydorn, Risø Report No. 326, Danish Atomic Energy Commission, Roskilde, 1976, p. 24.

31. J. Versieck, R. Cornelis, G. Lemey, and J. De Rudder, *Clin. Chem.*, *26*, 531 (1980).

32. L.-O. Plantin, in *Nuclear Activation Techniques in the Life Sciences* (Proceedings of a Symposium), International Atomic Energy Agency, Vienna, 1979, p. 321ff.

33. K. Kasperek, J. Kiem, G. V. Iyengar, and L. E. Feinendegen, *Sci. Total Environ.*, *17*, 133 (1981).

34. H. Nakahara, Y. Nagame, Y. Yoshizawa, H. Oda, S. Gotoh, and Y. Murahami, *J. Radioanal. Chem.*, *54*, 183 (1979).

35. H. J. M. Bowen, in *Critical Reviews in Analytical Chemistry* (B. Campbell and L. Meites, eds.), CRC Press, Florida, 1980, p. 127ff.

36. J. L. Carden and R. W. Fink, in *Zinc and Copper in Medicine*
     (Z. A. Karcioglu and R. M. Sarper, eds.), Charles C Thomas,
     Illinois, 1980, p. 3ff.

37. K. Heydorn, E. Damsgaard, N. A. Larsen, and B. Nielsen, in
     Nuclear Activation Techniques in the Life Sciences (Proceedings
     of a Symposium), International Atomic Energy Agency, Vienna,
     1979, p. 129ff.

38. R. Cornelis, A. Speecke, and J. Hoste, *Anal. Chim. Acta, 78,*
     317 (1975).

39. N. A. Larsen, H. Pakkenberg, E. Damsgaard, K. Heydorn, and
     S. Wold, *J. Neurol. Sci., 51,* 437 (1981).

40. D. Brune, G. F. Nordberg, P. O. Wester, and B. Bivered, in
     Nuclear Activation Techniques in the Life Sciences (Proceedings
     of a Symposium), International Atomic Energy Agency, Vienna,
     1979, p. 643ff.

41. S. Miyata, S. Nakamura, M. Toyoshima, Y. Hirata, M. Saito,
     M. Kameyama, R. Matushita, and M. Koyama, *Clin. Chim. Acta,
     106,* 235 (1980).

42. M. Yukawa, M. Suzuki-Yasumoto, K. Amano, and M. Terai, *Arch.
     Environ. Health, 35,* 36 (1980).

43. C. F. Clemente, L. Cigna Rossi, and G. P. Santaroni, *J. Radio-
     anal. Chem., 37,* 549 (1977).

44. K. Heydorn, E. Damsgaard, N. Horn, M. Mikkelsen, I. Tygstrup,
     S. Vestermark, and J. Weber, *Humangenetik, 29,* 171 (1975).

45. E. Damsgaard, K. Østergaard, and K. Heydorn, *J. Radioanal. Chem.,
     70,* 67 (1982).

46. J. Versieck, J. Hoste, F. Barbier, H. Steyaert, J. De Rudder,
     and H. Michels, *Clin. Chem., 24,* 303 (1978).

Chapter 10

DETERMINATION OF LEAD, CADMIUM, AND
MERCURY IN CLINICAL CHEMISTRY

Arne Jensen
Chemistry Department AD
Royal Danish School of Pharmacy
Copenhagen, Denmark

Jytte Molin Christensen
Danish National Institute of Occupational Health
Hellerup, Denmark

Poul Persson
Medi-Lab a.s.
Copenhagen, Denmark

1. CLINICAL CHEMICAL ASPECTS

When lead is inhaled the rates of deposition, retention, and absorption are highly variable, depending on particle size, physicochemical form of lead, and the efficiency of lung clearance mechanisms. About 30-45% of inhaled lead is absorbed [1], and daily lead intake with normal diet is in the industrial world about 100 µg of which 5-10% is absorbed.

The amount of lead bound to the body consists essentially of two compartments. Inorganic lead is found in bone (90% of the total content) and it has a half-time of about 20 years, and the concentration of lead in this compartment increases throughout the whole lifetime. The second and much smaller compartment (blood, soft tissue, etc.) has a half-time of only 20-30 days. Absorbed lead is excreted mostly in urine (80%) and by gastrointestinal secretion [1].

The most common symptom of acute lead poisoning is found as gastrointestinal colic. Chronic encephalopathy may result from prolonged lead absorption, but it may also be a residual effect of acute encephalopathy. Accidental and occupational exposure to alkyl lead compounds may develop into an acute encephalopathy which is different from the effects of inorganic lead compounds on the central nervous system. For differential diagnosis of lead it is essential to determine lead in blood and ALA and coproporphyrin in urine [1]. Furthermore, lead absorption causes an increase in the protoporphyrin content of the circulating erythrocytes. This protoporphyrin has been shown to be predominantly chelated with zinc [2].

Daily cadmium intake with normal diet is in the industrial world about 60 µg. It has been shown that about 5% of cadmium ingested by humans is absorbed, but a deficiency of calcium and iron may increase this amount. The cadmium absorbed is mainly stored in kidneys and liver, and the excretion in urine, which is extremely slow, less than 0.01% of the total amount in the body per day, corresponds to a biological half-time of more than 20 years in human beings [3].

Long-term exposure to cadmium via food and excessive exposure
to this element in small particles via inhalation can cause acute
lung disease and chronic renal disease. The appearance in the urine
of low-molecular-weight proteins, known as tubular proteinuria, is a
sign of chronic cadmium intoxication. It has been estimated that
long-term exposures with daily intakes of 300-400 μg cause renal
tubular dysfunction. Osteomalacia has been found both in industrially
exposed workers and in women exposed to excessive amounts of cadmium
in rice, a disease called Itai-Itai disease. Anemia and disturbed
liver function may also result from excessive cadmium exposure [3].

Mercury occurs as inorganic and organic compounds (mercury
vapor, mercury salts, and organic mercury compounds, e.g., phenyl
mercury compounds) [4].

The toxic properties of mercury vapor and organic bound mercury
are due to accumulation of this element in the brain causing neuro-
logical signs involving an unspecific psychoasthenic and vegetative
syndrome (micromercurialism). When a person is exposed to higher
levels of mercury tremor is seen, and this tremor is usually accompanied by severe behavioral and personality changes [4].

The hazards found in connection with long-term intake of food
containing methylmercury or occupational exposure to methylmercury
have been shown to result from the efficient absorption (90%) of
methylmercury in humans and the extremely long retention time (biological half-time of about 70 days) with an accumulation of methylmercury in the brain. Chronic poisoning results in degeneration and
athrophy of the sensory cerebral cortex [4].

2. DETERMINATION OF THE TOTAL CONTENTS
OF LEAD, CADMIUM, AND MERCURY*

2.1. Introduction

Schroeder and Nason [5] and also Kimura and Miller [6] have used a
spectrophotometric dithizone procedure for determination of lead and

---

*1 μg/liter = 4.8 nmol/liter (Pb), 8.9 nmol/liter (Cd), and 5.0
nmol/liter (Hg).

mercury, respectively. The spectrophotometric determination of lead, cadmium, and mercury in clinical chemistry is, however, not sufficiently sensitive and precise, and voltammetric methods or AAS are used for the determination of lead and cadmium or lead, cadmium, and mercury (cf. Sects. 2.2 and 2.3).

2.2. Anodic Stripping Voltammetry and Other Voltammetric Methods

The classical polarography is not in use any longer as far as clinical chemical analysis of the two elements mentioned is concerned [7].

Potentiometric stripping analysis has not been used for determination of lead and cadmium in clinical chemical analysis [8].

Anodic stripping voltammetry (ASV) is often called linear-potential sweep stripping chromamperometry [9]. The metal ions present in solution are in ASV precipitated (reduced) electrolytically on the cathode, which consists of impregnated graphite. The graphite has been covered electrolytically with a colloidal layer of mercury. The ASV method consists of two steps. In *the first step* a plating or reduction of the metal ions takes place since the electrode potential is kept sufficiently negative, and the magnitude of the potential depends on the metal ions of interest. The electrode potential is varied in *the second step*, with a constant rate and in a positive, i.e., *anodic* direction.

Stripping is found in several different forms. The most common analysis form is linear stripping, but the fast and continuous change of the potential gives rise to a pronounced background current. The differential pulse technique and the stepwise stripping reduce the background current, but these two techniques extend the time of analysis [10].

Morrell and Giridhar [11] described a rapid microscale procedure for blood lead determination by ASV (80-1000 μg/liter). Results corre-

lated well with those obtained by AAS. The procedure involved use of
a metal-exchange reagent which rapidly released the bound lead from
its micromolecular binding sites thus eliminating a long and cumber-
some perchloric acid digestion procedure and a risk of contamination.
The active component of the metal-exchange reagent was chromium(III)
chloride. Reproducibility and analytical recovery of added lead were
excellent and loss by sample preparation or contamination was eliminated. The results show that ASV is a reliable, sensitive micro-
method for determination of blood lead.

Ferren [12] published limited data on the cadmium and lead
content of nails and hair, as well as in environmental samples. All
samples were prepared for the analysis by carrying out a wet oxidation
by refluxing at 220°C for 20 min with 70% perchloric acid, 70% nitric
acid, and 98% sulfuric acid at a ratio of 24:24:1. The cadmium content of nails and hair was found to be 2 and 1 mg/kg, respectively,
and the lead content to be 26 and 40 mg/kg, respectively; the content
shown was indicated to be typical values in samples from subjects
living in Staten Island, New York.

Christensen and Angelo [13] described a rapid, sensitive ASV
method for measuring concentrations of cadmium in blood. An advantage of the method was the minimal sample preparation required when
a metal-exchange reagent was used (cf. the work by Morrell and
Giridhar [11]). The reference interval for blood cadmium was 0.56-
6.72 μg/liter for nonsmokers and 0.56-9.52 μg/liter for cigarette
smokers. The sensitivity of this ASV method for Cd in blood was
0.45 μg/liter (twice the standard deviation). It was shown that
magnesium, iron, zinc, copper, lead, sodium, potassium, chloride,
phosphate, and bicarbonate did not interfere when the determination
of Cd in blood was carried out. EDTA has for many years been used
as a chelating agent in heavy-metal poisoning, and EDTA reduces the
recovery of Cd in blood. However, addition of nickel chloride to
the sample reverses the inhibition. The ASV method gave values in
reasonable agreement with AAS, and a Danish national quality inter-
comparison program showed that the precision for Cd was satisfactory.

2.3. Atomic Absorption Spectroscopy and Related Methods

2.3.1. Lead and Cadmium

Both lead and cadmium present in biological tissues, blood, urine, etc., have been determined using flame AAS with a suitable pretreatment of the samples to be analyzed. It is, however, rather unusual now to use this classical method for the determination of these two elements in clinical chemical analysis.

Sperling [14] published methods for the determination of cadmium in environmental samples by means of flameless AAS. The method described in the publication consisted of an acid digestion procedure followed by neutralization, extraction, and flameless AAS measurement. The main characteristics were:

1. Digestion took place in small 1.5-ml tubes of quartz, polyvinyldifluoride, or polypropylene.
2. Neutralization was not performed by titration but only by an excess of saturated $NaHCO_3$ solution.

Sperling and Bahr [15,16] showed that background levels from tissues of nonexposed or only slightly exposed marine organisms were very difficult to measure. Among these blood and urine presented extreme difficulties because of their high content of both organic materials and salts. The standard addition method led to wrong results, and the effect observed was interpreted as an effect of a concurrence between other heavy metals in a considerable concentration (iron was suggested) and cadmium during extraction with APDC. However, an increase of the APDC concentration in the extractant led to results which could be referred to aqueous standards directly. It was shown that the heavy metal could not be extracted completely without prior digestion of the liquid sample of blood or urine.

Schramel et al. [17] invented a new device for ashing of biological material under pressure, and the range of application was tested with different standard reference materials for cadmium, lead, and several other elements. Lead and cadmium were determined using both flameless AAS and ICP spectroscopy. Only the AAS method was useful because it was shown that ICP spectroscopy was relatively insensitive to Pb and Cd, although not to the other elements examined, i.e., Cu, Mn, Fe, and Zn.

Carter and Yeoman [18] described a rapid method for digesting organic material (in casu: whole blood) prior to AAS using low-temperature ashing. A vacuum of 0.133 kPa or less was produced above the sample (10 μl whole blood and 50 μl deionized, distilled water) after having dried this sample at 110°C for 2 min. A gas mixture of oxygen and carbon tetrafluoride was then allowed to bleed into the system, and a radiofrequency power was tuned to give maximum forward power (100 W) and minimum reflected power (less than 5 W). It was found that a 1 + 1 mixture of the two gases provided the optimum atmosphere for low-temperature ashing. By this method it was possible to achieve determinations of Cd in blood within the range 0.2-20 μg/liter. The calculated limit of determination, using blood added 5 μg/liter of Cd, was 0.33 μg/liter, and an apparent loss of Cd during the low-temperature ashing was not detected.

Subramanian and Meranger [19] described a rapid, simple, and sensitive in situ $(NH_4)_2HPO_4$-Triton X-100 electrothermal AAS procedure for the determination of Pb and Cd in whole human blood. The heparinized human whole blood is diluted fivefold with an aqueous solution of $(NH_4)_2HPO_4$ and Triton X-100, and a 10-μl aliquot of this solution is injected into a pyrocoated graphite tube. The amount of Pb and Cd in the blood sample is calculated by comparison to linear working curves prepared from aqueous standards in $(NH_4)_2HPO_4$ + Triton X-100. The method is free of matrix effects and there is no need to use the method of standard addition or matrix-matched calibration curves. Analysis of whole blood is desirable because more than 95% of Pb [20] and Cd [21] is in the erythrocytes. The linear range, detection limit (3 SD of blank), and sensitivity for Cd and Pb in blood, respectively, are (in μg/liter) 0.0-3.0, 0.10, 0.05, and 0.0-5.0, 2.0, and 1.5.

Pleban et al. [22] used a method with a background correction system based on the polarization characteristics of the Zeeman effect.

Zeeman effect AAS offers accurate and reproducible background correction during atomization for broad-band molecular absorption, light scattering, and wavelength-dependent radiation. This AAS method with Zeeman background corrector was used to measure lead and cadmium (as well as copper and manganese) in a nitric acid digest of lyophilized human kidney cortex. Average analytical recoveries for lead and cadmium were 96 and 95%, respectively. The assay was linear to 200 µg/liter for lead and to 75 µg/liter for cadmium. Mean trace metal concentrations in human kidney cortex, given as mg/kg of lyophilized tissue, were found to be 2.25 ± 1.20 for lead and 139 ± 88 for cadmium in 30 kidney cortex specimens. This value for cadmium agrees well with other reported values [23,24].

Sumino et al. [25] developed a new flameless AAS method with an optical microscope and laser oscillator for determination of metals. The method was used for microlocalization of cadmium in human kidney cortex.

Blacklock and Sadler [26] described a semiquantitative screening method, where the sample, after complexation and extraction, was analyzed by direct current arc emission spectroscopy and the spectra were recorded on a photographic plate. The method has been used for determination of lead, cadmium, mercury, and a few other elements in biological material.

Bruhn and Navarrete [27] described a matrix modification procedure for the direct determination of Cd in urine by electrothermal AAS. Matrix modification with ammonium nitrate, ammonium dihydrogenphosphate, and Triton X-100 proved suitable. Optimization of the graphite furnace parameters allowed cadmium to be quantified at 800°C. The sensitivity was 0.1 µg/liter. Urinary cadmium levels of 0.4-1.8 µg/liter were measured in five occupational unexposed persons.

2.3.2. Mercury (Nonflame Method at Room Temperature)

Hatch and Ott [28] described an extremely sensitive and accurate method for the determination of mercury. The detection limit in solution was 1.0 µg/liter. This method has been applied to soil

samples containing organic materials. The sample was brought in
solution by an oxidizing acid attack. Mercury was then reduced to
the elemental state by means of tin(II) sulfate solution. The mercury vapor was passed through a quartz absorption cell where it was
measured quantitatively. The procedure was free from interferences
due to organic matter or other volatile constituents of the sample.
Large amounts of easily reducible elements (e.g., Cu) must not be
present in the solution, but the method could be used in samples
with, for example, Ni and Co. The procedure outlined was the first
method published for determination of Hg by a nonflame method at room
temperature. The results obtained by Hatch and Ott were compared
with NAA and AAS methods, and with dithizone spectrophotometry.

Magos [29] described a simple method for the determination of
total mercury in biological samples contaminated with inorganic mercury and methylmercury. The method was based on the rapid conversion
of organomercurials first into inorganic mercury and then into atomic
mercury suitable for aspiration through the gas cell by a combined
tin(II) chloride-cadmium chloride reagent as reductor. It was found
that if 100 mg of tin(II) chloride alone was added instead of the
tin(II) chloride-cadmium chloride reagent, only the release of inorganic mercury influenced the peak deflection of the recorder, this
permitting the selective determination of inorganic mercury and
then, after reacidification of the reaction mixture, methylmercury,
by adding tin(II) chloride-cadmium chloride reagent and sodium hydroxide. When total mercury and inorganic mercury were determined sepa-
rately, the difference between results gave the methylmercury content
of the sample. The method was shown not to be specific in the sense
that phenylmercury behaved like methylmercury and about 30% of the
mercury was released from ethylmercury even when tin(II) chloride
was used. However, in food that has *not* been directly contaminated
with phenylmercury or ethylmercury, the mercury contaminant was
always either inorganic mercury or methylmercury. The method suggested by Magos has been used for many different biological materials,
e.g., blood, kidney, liver, brain, tuna fish and fishmeal, and in

TABLE 1

Summary of Methods and Conditions

Element	Method	Detection limit	Sensitivity	Standard deviation	Pretreatment of sample and analytical range	Biological material	Ref.
Pb	ASV			25 µg/liter	Direct determination 80-1000 µg/liter	Blood	[11]
Pb Cd	ASV				Wet digestion	Nails and hair	[12]
Cd	ASV		0.45 µg/liter	0.23 µg/liter	Direct determination 0.56-6.72 µg/liter (non smokers) 0.56-9.52 µg/liter (smokers)	Blood	[13]
Cd	Flameless AAS	1.2 µg/kg			Wet digestion and extraction 0.20 µg/kg	Marine organisms	[14]
Cd	Flameless AAS				Wet digestion and extraction 0-6.0 µg/liter (blood) 0-4.0 µg/liter (urine	Blood and urine	[15,16]
Pb Cd	Flameless AAS				Wet ashing under pressure 0-50 mg/kg (Pb) 0-3 mg/kg (Cd)	Bovine liver	[17]
Cd	AAS	0.33 µg/liter			Low temp. dry ashing under vacuum + O_2 0.2-20 µg/liter	Blood	[18]
Pb Cd	Flameless AAS	2.0 µg/liter 0.10 µg/liter	1.5 µg/liter 0.05 µg/liter	0.7 µg/liter 0.03 µg/liter	Direct determination 0.0-50.0 µg/liter (Pb) 0.0-3.0 µg/liter (Cd)	Whole blood	[19]

Element	Method	Detection limit		Sample preparation	Sample	Ref.
Pb Cd	Polarized Zeeman effect flameless AAS	1.2 µg/liter 1.2 µg/liter	0.6 µg/liter 0.6 µg/liter	Wet digestion 0-200 µg/liter (Pb) 0-75 µg/liter (Cd)	Lyophilized human kidney cortex	[22]
Cd	Laser oscillator + flameless AAS			Direct determination	Human kidney cortex	[25]
Pb Cd Hg	Arc emission spectroscopy			Wet digestion, complexation and extraction	Tissue	[26]
Cd	Flameless AAS	0.1 µg/liter	0.1 µg/liter	Direct determination 0.0-2.0 µg/liter	Urine	[27]
Hg (tot.)	Flameless spectrophotom.	1.0 µg/liter		Acidic digestion	Organic materials from soil samples	[28]
Hg (inorg.) Hg (tot.)	Flameless spectrophotom.			Basic digestion	Blood and tissue	[29]
Hg (tot.)	Flameless spectrophotom.			Acidic digestion	Blood, urine and fish tissue	[30]
Hg (inorg.) Hg (tot.)	Flameless spectrophotom.	2.4 µg/liter Hg (tot.) urine 3.8 µg/liter Hg (tot.) blood		Urine: direct determin. Blood: acid digestion	Urine Hg (tot.) and blood	[31]
Hg (tot.)	Flameless spectrophotom.			Basic permanganate oxidation and sulfuric acid digestion	Urine and blood. Other biological materials	[34]

concentrations as low as 0.05 μg of mercury in 5 ml of 20% tuna fish
or 10% liver homogenate, but the error could in this case be 25%.

Skare [30] dealt with the extended application of the automated
mercury analysis equipment to fish and blood samples which are of
great interest in the public health area. With the standard procedures recommended it was possible to determine mercury concentrations
in blood down to about the normal values for unexposed persons, i.e.,
about 5 μg/kg, and mercury in fish down to the 100 μg/kg level. Com-
parative studies between the methods published and the methods of NAA
and GC showed good agreement.

Coyle and Hartley [31] took advantage of Magos reagent [29,32,
33], in that it permits the analyst to discriminate between organic
and inorganic mercury. The procedure with a fast and simple pretreatment step allows the measurement of mercury in both inorganic and
organic forms in blood by an automated technique. It was possible
to estimate concentrations of total mercury in urine and blood down
to 2.4 and 3.8 μg/liter respectively. The overall mean percentage
recovery was 97-100% for urine and blood.

Chapman and Dale [34] used alkaline permanganate in the preparation of biological materials for the determination of mercury by AAS.
This treatment more than halved the processing time required by the
acid permanganate procedure. Dissolution was completed by addition
of sulfuric acid and excess of oxidant, then reduced with oxalic acid.
The stable mercury(I) oxalate complex was then reduced by tin(II).
The homogenization step allowed representative subsampling. The procedure was applicable to urine, blood, and other biological materials,
and mercury in the amount of 1.1 μg/liter could be easily detected.

2.4. Determination of Ion Activity

The activity of lead and cadmium ions has been determined for one to
two decades by using polycrystalline ion-selective electrodes [35],
but in blood lead and cadmium are protein-bound, i.e., determination
of the activity is of no interest.

3. SUMMARY OF METHODS AND CONDITIONS

An overview of the methods and the conditions used in clinical chemistry for the determination of lead, cadmium, and mercury is given in Table 1.

ABBREVIATIONS AND DEFINITIONS

AAS	atomic absorption spectroscopy
ALA	δ-aminolevulinic acid
APDC	ammonium pyrrolidine dithiocarbaminate
ASV	anodic stripping voltammetry
Detection limit	concentration that gives a signal-to-noise ratio of 2
GC	gas chromatography
ICP	inductively coupled plasma
NAA	neutron activation analysis
SD	standard deviation
Sensitivity	concentration that produces an absorbance of 0.0044 (1% absorption)

REFERENCES

1. K. Tsuchiya, in *Handbook on the Toxicology of Metals* (L. Friberg, G. F. Nordberg, and V. B. Vouk, eds.), Elsevier/North Holland, 1979, p. 451.

2. P. Grandjean and J. Lintrup, *Scand. J. Clin. Lab. Invest.*, *38*, 669 (1978).

3. L. Friberg, T. Kjellström, G. F. Nordberg, and M. Piscator, in *Handbook on the Toxicology of Metals* (L. Friberg, G. F. Nordberg, and V. B. Vouk, eds.), Elsevier/North Holland, 1979, p. 355.

4. M. Berlin, in *Handbook on the Toxicology of Metals* (L. Friberg, G. F. Nordberg, and V. B. Vouk, eds.), Elsevier/North Holland, 1979, p. 503.

5. H. A. Schroeder and A. P. Nason, *Clin. Chem.*, *17*(6), 461 (1971).

6. Y. Kimura and V. L. Miller, *Anal. Chim. Acta*, *27*, 325 (1962).

7. O. Siggaard-Andersen, in *Fundamentals of Clinical Chemistry* (N. W. Tietz, ed.), W. B. Saunders, London, 1976, p. 147.

8. D. Jagner and K. Årén, *Anal. Chim. Acta, 107,* 29 (1979).

9. H. H. Willard, L. L. Merritt, Jr., J. A. Dean, and F. A. Settle, Jr., *Instrumental Methods of Analysis,* 6th ed., D. Van Nostrand Company, London, 1981, p. 715.

10. J. Molin Christensen, private communication.

11. G. Morrell and G. Giridhar, *Clin. Chem., 22*(2), 221 (1976).

12. W. P. Ferren, *Int. Lab. (Sept./Oct.),* 55 (1978).

13. J. Molin Christensen and H. Angelo, *Scand. J. Clin. Lab. Invest., 38,* 655 (1978).

14. K.-R. Sperling, *Fresenius' Z. Anal. Chem., 299,* 103 (1979).

15. K.-R. Sperling and B. Bahr, *Fresenius' Z. Anal. Chem., 301,* 29 (1980).

16. K.-R. Sperling and B. Bahr, *Fresenius' Z. Anal. Chem., 301,* 31 (1980).

17. P. Schramel, A. Wolf, R. Seif, and B.-J. Klose, *Fresenius' Z. Anal. Chem., 302,* 62 (1980).

18. G. F. Carter and W. B. Yeoman, *Analyst, 105,* 295 (1980).

19. K. S. Subramanian and J. C. Meranger, *Clin. Chem., 27*(11), 1866 (1981).

20. C. N. Ong and W. R. Lee, *Br. J. Ind. Med., 37,* 70 (1980). Cit. from Ref. 19.

21. L. Friberg, M. Piscator, G. F. Nordberg, and T. Kjellström, *Cadmium in the Environment,* CRC Press, Cleveland, Ohio, 1974. Cit. from Ref. 19.

22. P. A. Pleban, J. Kerkay, and K. H. Pearson, *Clin. Chem., 27*(1), 68 (1981).

23. C.-G. Elinder, T. Kjellström, B. Lind, L. Linnman, and L. Friberg, *Arch. Environ. Health, 31,* 292 (1976). Cit. from Ref. 22.

24. Cf. Ref. 3, p. 374.

25. K. Sumino, R. Yamamoto, F. Hatayama, S. Kitamura, and H. Itoh, *Anal. Chem., 52,* 1064 (1980).

26. E. C. Blacklock and P. A. Sadler, *Clin. Chim. Acta, 113,* 87 (1981).

27. C. F. Bruhn and G. A. Navarette, *Anal. Chim. Acta, 130,* 209 (1981).

28. W. R. Hatch and W. L. Ott, *Anal. Chem., 40,* 2085 (1968).

29. L. Magos, *Analyst, 96,* 847 (1971).

30. I. Skare, *Analyst, 97,* 148 (1972).

31. P. Coyle and T. Hartley, *Anal. Chem.*, *53*, 354 (1981).

32. M. R. Greenwood, P. Dhahir, and T. W. Clarkson, *J. Anal. Toxicol.*, *1*, 265 (1977). Cit. from Ref. 31.

33. P. N. Kubasik, H. E. Sine, and M. T. Volosin, *Clin. Chem.*, *18*, 1326 (1972). Cit. from Ref. 31.

34. J. F. Chapman and L. S. Dale, *Anal. Chim. Acta*, *134*, 379 (1982).

35. Cf. Ref. 9, p. 643.

Chapter 11

DETERMINATION OF CHROMIUM IN URINE AND BLOOD

Ole Jøns and Arne Jensen
Chemistry Department AD
Royal Danish School of Pharmacy
Copenhagen, Denmark

Poul Persson
Medi-Lab a.s.
Copenhagen, Denmark

1. CLINICAL CHEMICAL ASPECTS

Chromium(III) is an essential element in humans and plays an important role in insulin metabolism as part of the glucose tolerance factor. The daily requirement of chromium in human nutrition has not been established.

Chromium has also been implicated in the control of cholesterol and lipid biosynthesis, and many papers dealing with the presumed

relation of the element to atherosclerosis and diabetes [1] have
appeared.

The difficulty of studying the biological role of chromium is
underlined by the fact that it was found to be effective at doses as
low as 0.01 µg/kg of body weight.

It has been indicated that chromium(VI) can easily be absorbed
and that this oxidation state is carcinogenic. Workers in contact
with chromium(VI) have frequently developed lung cancer and/or dermatitis of the hands [2].

It is, therefore, also of interest to determine the concentration of chromium(III or VI) in blood, serum, or plasma.

2. DETERMINATION OF THE TOTAL CONTENTS
OF CHROMIUM IN URINE OR BLOOD*

2.1. Atomic Absorption Spectroscopy (Classical and Flameless Methods)

Much effort has been devoted to the determination of chromium in
biological samples. A large body of analytical data has accumulated,
but a deep confusion remains as to the actual concentration in plasma
or serum (and urine) of healthy individuals as well as of patients
suffering from some kind of intoxication of chromium(VI). Widely
divergent values have been reported for plasma or serum chromium concentration, varying from 0.14 to 185 µg/liter [3].

Therefore, the accurate determination of chromium in biological
specimens in general is a severe challenge, as can also be seen from
the results of interlaboratory comparisons of reference materials [4].
Two main causes have been postulated to explain the existing discrep-
ancies:

1. Volatilization during sample drying and/or dry ashing, and

2. Contamination during sample collecting and handling
Cumulative evidence has indicated that the first explanation is not
correct. In contrast, the second explanation, viz. inadequacy of

---

*1 µg/liter = 1.9 x 10^{-5} mmol/liter (Cr).

sample collecting and handling, is a major source of error, e.g., adequate blood collection has been proven to be of vital importance in ensuring reliable results.

The remaining part of this section will in chronological order cover a representative part of the literature dealing with the determination of chromium in urine and blood by AAS methods.

The widely used techniques for chromium determinations in biological materials, classical AAS [5] and AES [6], were refined and improved in sensitivity [7], but these classical techniques are not in use any longer for the determination of chromium in the materials relevant for clinical chemical analysis. A several hundredfold improvement in absolute sensitivity for chromium has been made possible by the appearance of the electrically heated graphite atomizer. With this atomizer, the sample is introduced directly into the tube, and provision is made for programming the temperature of the tube for sequential operations of sample drying, ashing, and atomization. Beside the greater sensitivity afforded by increased efficiency of atomization, other potential advantages of the graphite tube atomizer for the analysis of chromium in biological fluids lie in the small sample volume required (5-50 µl) and the potential for analyzing samples containing organic matter without pretreatment.

Davidson and Secrest [8] carried out a study with the purpose of evaluating the graphite tube atomizer as a practical means of performing the quantitative determination of chromium in blood plasma and urine. For routine analysis of plasma, the method was found to be effective with either preashed samples or direct-sample analysis of very small amounts of the order of 20-200 µl. The method has also been used for the determination of chromium in urine, whole blood, and other tissue samples. Chromium concentrations in fasting plasma from seven subjects ranged from 3.10 to 7.19 µg/liter with a mean of 5.07 µg/liter. It was shown in a brief study that the absorption line for chromium (357.9 nm) was not affected by the presence of 500- to 5000-fold excess of various cations and anions. The calibration curve for chromium in the presence of these ions was identical with the curve obtained with chromium standard solutions, and similar

experiments showed 100% recovery. It was shown by Davidson and
Secrest [8] that pretreatment of the sample by wet ashing should be
carried out because (1) chemical interferences due to sample composition were avoided, (2) there was general applicability to the
variety of biological materials, and (3) considerable reduction of
instrument analysis time with wet-ashed samples has occurred. Pretreatment of the sample by wet ashing, however, necessitated a blank
correction that amounted to about 20% of the normal content of plasma
and urine samples. Since the blank determinations were highly reproducible, it was concluded that acid leaching of chromium from the
glass of digestion vials contributed significantly to the blank.
This was verified by coating the vials with silicone.

Grafflage et al. [9] also used flameless AAS and a wet-ashing
technique in their studies on the determination of chromium in serum.
The authors have optimized both the wet ashing and the atomization
temperature and have used Suprapure reagents from Merck. Using this
method at 357.9 nm, the concentrations of chromium were measured in
the sera of 50 apparently healthy persons. The concentration of the
element showed a normal log distribution and the concentrations found
were 0.23-1.90 µg/liter with an arithmetic mean value of 0.73 µg/liter.
This value agreed well with the result of Mertz et al. [10], who used
emission spectral analysis, and with the value of Kasperek et al. [11],
who used the NAA technique and obtained a very small standard deviation.

Kayne et al. [12] used a flameless AAS method for the determination of chromium in serum and urine. The data obtained indicate that
the method used allows routine determinations of as little as 0.1 µg/
liter in samples of biological fluids. The principal reason for this
sensitivity is the improved analytical instrumentation as well as the
sample procedure, which routinely results in *blank* values of less
than 0.1 µg/liter. The sample procedure is different from the procedures generally used since it has been found that a wet-ashing technique is unnecessary. The mean value determined by these authors was
0.14 µg/liter, in excellent agreement with the value of 0.16 µg/liter
recently reported in an NAA study [13].

Guthrie et al. [14] have also investigated factors influencing the determination of Cr in urine by flameless AAS. Tests showed that the conventional deuterium background corrector was inadequate. It was demonstrated that sensitivity increased about threefold with pyrolytically coated graphite tubes. It was furthermore concluded that an alternative system allowing for adequate background correction is needed at 357.9 nm. The alternatives seemed to include the polarized Zeeman effect [15] (vide infra), a continuum source wavelength modulated, echelle AAS system [16,17], or the modification of existing D_2 systems by substituting for the D_2 lamp a continuum source with greater intensity at the Cr wavelength. Preliminary studies carried out with an echelle AAS system was found to be satisfactory for the determination of Cr in urine, and the concentration was found to be less than 1 µg/liter.

Routh [18] evaluated the analytical parameters for the determination of chromium in urine by flameless AAS. The reduction of and compensation for nonspecific absorption interferences has been shown to be the primary factor limiting analytical accuracy. Reduction of nonspecific absorption to an acceptable level was established by means of a hydrogen diffusion flame as a supplement to the inert gas (nitrogen) sheathing the graphite absorption cell. Simultaneous continuum source background correction by means of a deuterium arc lamp was shown to be adequate for nonspecific absorption compensation. The typical limit of detection for chromium in the urine matrix (20-µl sample) was found to be 0.2 µg/liter.

Kumpulainen [19] used a flameless AAS method and compared the results of low-temperature ashing and dry ashing at 500°C for the analysis of chromium in urine (and human milk). Dry ashing gives superior accuracy. The importance of contamination control, background correction, and temperature control are stressed. It was found that urine (and milk) contain 1 µg/liter. The causes of the losses of chromium in the low-temperature ashing have not been found, but it is suggested that the superior accuracy using the dry-ashing procedure might be due to the better temperature control system of the AAS instrument used.

Graf-Harsanyi and Langmyhr [20] used flameless AAS methods to
determine the total content of chromium in a lyophilized reference
sample of animal serum. Lyophilized samples were wet-ashed in platinum dishes and polytetrafluoroethylene-lined aluminum bombs. The
concentration in the solid sample was found to be 61 µg/kg. It was
shown that chloride (and presumably other halides) is removed by
evaporation on subsequent heating, when the analysis is based on
matrix modification by adding (to the solid or liquid sample deposited in the graphite tube) a solution of nitric acid, sulfuric acid,
or ammonium nitrate. The matrix modification mentioned gave the best
results, thus avoiding the possible loss of chromium as the halide
during the pyrolysis and atomization steps. The authors showed,
furthermore, that the distribution of chromium among the serum proteins, as studied by gel filtration and flameless AAS, was found to
be concentrated in the high- and low-molecular-weight protein fractions.

Wang [21] determined the concentration of chromium(VI) in
aqueous solutions as a thiosemicarbazide complex by solvent extraction into MIBK and a classical AAS method, but the method was not
used in connection with clinical chemical analysis. Thiosemicarbazide
was chosen for the investigation as a chelating agent for chromium
because it has both a hard-base site (nitrogen) and a soft-base site
(sulfur), and is well known as a useful chelating agent for transition metal ions in general.

Veillon et al. [22] showed by means of a ^{51}Cr tracer technique
that a considerable amount of chromium from rat urine is irreversibly
retained in the AAS graphite furnace tubes upon atomization. Both
atomization temperature and sample matrix were found to be very
influencial in the amount retained, and pyrolytically coated tubes
retained less Cr than uncoated tubes. Considerable errors are therefore likely if the method of additions is not used. The amount of
Cr retained is presumably due to carbide formation. It was furthermore demonstrated (Table 1) that for charring temperatures of 1300°C
or less, no loss of Cr was observed, and above this temperature the

TABLE 1

Effect of Char Temperature and Time on Cr
Retention Within the Graphite Tube[a]

Indicated temp. (°C)	Char time (sec)	^{51}Cr retained (%)
1000-1300	30-90	100
1400	30	97
1400	60	94
1500	30	95
1500	60	92
1600	30	87
1600	60	84
1700	30	69
1700	60	65

[a]Uncoated tubes. Each time/temperature pair
represents the mean of two tubes.

Source: Reprinted by permission from Ref. 22,
© 1980 American Chemical Society.

Cr loss becomes progressively greater. Table 2 shows the effect of
atomization temperature and time on Cr retention within the graphite
tube. The authors conclude that "it is not surprising that conflict-
ing and erroneous urinary Cr values have appeared in the literature,"
and that "these retention data emphasize the need for precise temperature control in the furnace." No evidence of loss of chromium by
volatilization was found during analytical or sample preparatory
steps.

Veillon et al. [23] have continued their work on direct determination of chromium in urine by flameless AAS and have used the
method of standard additions to compensate for changes in sensitivity
as the furnace tube ages as well as for the widely used different
matrices encountered in urine samples. Agreement with independent
methods has been evaluated.

TABLE 2

Effect of Atomization Temperature and Time on
Cr Retention Within the Graphite Tube[a]

Indicated temp. (°C)	^{51}Cr retained (%)	
	Atomize 9 sec	Atomize 15 sec
2300	69	50
2400	57	40
2500	52	35
2600	48	33
2700	45	

[a]Uncoated tubes. Each measurement represents
the mean of two tubes.

Source: Reprinted by permission from Ref. 22,
© 1980 American Chemical Society.

Fernandez et al. [24] compared Zeeman effect AAS to conventional AAS with respect to sensitivity, analytical range, and background correction performance. The authors employed a versatile
Zeeman test instrument capable of both flame and furnace operation,
and sensitivities were compared for a total of 44 elements including
chromium. The sensitivity was smaller employing the Zeeman effect
instrument, but the background correction was much better in a
variety of difficult sample matrices using the graphite furnace.
The authors' work did not deal with background problems of the
clinical chemical analysis of chromium.

2.2. Spectrophotometry

Urone and Anders [25] described a rapid, spectrophotometric method
for the determination of chromium in human blood, tissues, and urine.
The sensitivity of the method was shown to be 5 μg/liter of Cr.
Samples were ashed in borosilicate glass beakers, using a wet-ashing

method for urine and dry-ashing methods for blood and tissues. The samples were oxidized and the best method for estimating minute amounts of Cr(VI) was (back in 1950) a spectrophotometric one, where diphenylcarbazide was used as the reagent.

Bryson and Goodall [26] distinctly improved the spectrophoto-metric determination of chromium in liver tissue digests with diphenyl carbazide. In this investigation, the authors examined chromium-spiked mouse liver tissue samples, which were wet-ashed. Ce(IV) oxidized Cr(III) completely at room temperature, the rate of oxida-tion being dependent on pH and matrix composition, and it was estab-lished that a pH of 2 (to prevent hydrolysis, etc.) was a convenient acidity for the analysis. Standard Cr(III) and Cr(VI) solutions and mineralized liver samples containing standard additions of Cr(III) or Cr(VI) could, with the modified technique, be measured in the range 20.0-4.00 x 10^3 µg/liter with a spectrophotometric sensitivity of 1.18 µg/liter and with no loss of chromium occurring during the digestion stage.

Table 3 contains a summary of the methods and conditions used in the determination of chromium in clinical chemistry.

ABBREVIATIONS AND DEFINITIONS

AAS	atomic absorption spectroscopy
AES	atomic emission spectroscopy
Detection limit	concentration that gives a signal-to-noise ratio of 2
MIBK	methyl isobutyl ketone
NAA	neutron activation analysis
Sensitivity	concentration that produces an absorbance of 0.0044 (1% absorption)

TABLE 3

Summary of Methods and Conditions

Method	Detection limit	Sensitivity	Standard deviation	Pretreatment of sample and analytical range	Biological material	Ref.
Flameless AAS	2 pg	44 pg		Wet ashing 3.10-7.19 µg/liter	Urine Whole blood	[8]
Flameless AAS				Wet ashing 0.23-1.90 µg/liter	Serum	[9]
Flameless AAS		0.1 µg/liter		Direct determ. 0.14 µg/liter	Urine Serum	[12]
Flameless AAS				<1 µg/liter	Urine	[14]
Flameless AAS	0.2 µg/liter			Direct determ.	Urine	[18]
Flameless AAS				Dry ashing 1 µg/liter	Urine	[19]
Flameless AAS				Wet ashing 61 µg/kg	Serum	[20]
AAS				Extraction	Aqueous solutions	[21]
Flameless AAS + tracer technique				Direct determ. or wet ashing retention data	Rat urine	[22]
Flameless AAS	0.03-0.04 µg/liter		0.012 µg/liter	Direct determ.	Urine	[23]
Spectrophotometry		5 µg/liter		Wet ashing (urine), dry ashing (blood, tissues)	Blood Tissues Urine	[25]
Spectrophotometry		1.18 µg/liter		Wet ashing 20.0 µg/liter- 4.00 mg/liter	Liver tissue	[26]

REFERENCES

1. D. Shapcott and J. Hubert (eds.), *Chromium Nutrition and Metabolism*, Vol. 2, Elsevier/North Holland, Amsterdam, 1979.

2. S. Langård and T. Norseth, in *Handbook on the Toxicology of Metals* (L. Friberg, G. F. Nordberg, and V. B. Vouk, eds.), Elsevier/North Holland, 1979, p. 383.

3. J. Versieck and R. Cornelis, *Anal. Chim. Acta, 116,* 217 (1980).

4. R. M. Parr, *J. Radioanal. Chem., 39,* 421 (1977).

5. F. J. Feldman, E. C. Knoblock, and W. C. Purdy, *Anal. Chim. Acta, 38,* 489 (1967).

6. I. H. Tipton, M. J. Cook, R. L. Steiner, C. A. Boye, H. R. Perry, and H. A. Schroeder, *Health Phys., 9,* 89 (1963). Cit. from Ref. 8.

7. K. M. Hambidge, *Anal. Chem., 43,* 103 (1971).

8. I. W. F. Davidson and W. L. Secrest, *Anal. Chem., 44,* 1808 (1972).

9. B. Grafflage, G. Buttgereit, W. Kübler, and H. M. Mertens, *Z. Klin. Chem. Klin. Biochem., 12,* 287 (1974).

10. D. P. Mertz, R. Koschnick, G. Wilk, and K. Pfeilsticker, *Z. Klin. Chem. Klin. Biochem., 6,* 171 (1968). Cit. from Ref. 9.

11. K. Kasperek, G. V. Iyengar, J. Kiem, H. Borberg, and L. E. Feinendegen, *Clin. Chem., 25,* 711 (1979).

12. F. Kayne, G. Komar, H. Laboda, and R. E. Vanderlinde, *Clin. Chem., 24,* 2151 (1978).

13. J. Versieck, J. Hoste, F. Barbier, H. Steyaert, J. De Rudder, and H. Michels, *Clin. Chem., 24,* 303 (1978).

14. B. E. Guthrie, W. R. Wolf, and C. Veillon, *Anal. Chem., 50,* 1900 (1978).

15. H. Koizumi, K. Yasuda, and M. Katayama, *Anal. Chem., 49,* 1106 (1977).

16. A. T. Zander, T. C. O'Haver, and P. N. Keliher, *Anal. Chem., 48,* 1166 (1976).

17. J. M. Harnly and T. C. O'Haver, *Anal. Chem., 49,* 2187 (1977).

18. M. W. Routh, *Anal. Chem., 52,* 182 (1980).

19. J. Kumpulainen, *Anal. Chim. Acta, 113,* 355 (1980).

20. E. Graf-Harsanyi and F. J. Langmyhr, *Anal. Chim. Acta, 116,* 105 (1980).

21. W.-J. Wang, *Anal. Chim. Acta, 119,* 157 (1980).

22. C. Veillon, B. Guthrie, and W. R. Wolf, *Anal. Chem.*, *52*, 457 (1980).

23. C. Veillon, K. Y. Patterson, and N. A. Bryden, *Anal. Chim. Acta*, *136*, 233 (1982).

24. F. J. Fernandez, S. A. Myers, and W. Slavin, *Anal. Chem.*, *52*, 741 (1980).

25. P. F. Urone and H. K. Anders, *Anal. Chem.*, *22*, 1317 (1950).

26. W. G. Bryson and C. M. Goodall, *Anal. Chim. Acta*, *124*, 391 (1981).

Chapter 12

DETERMINATION OF ALUMINUM IN CLINICAL CHEMISTRY

Arne Jensen
Chemistry Institute AD
Royal Danish School of Pharmacy
Copenhagen, Denmark

Erik Riber and Poul Persson
Medi-Lab a.s.
Copenhagen, Denmark

1. CLINICAL CHEMICAL ASPECTS

Aluminum-containing antacids are commonly used in therapy for peptic ulcer, and it is common to prescribe oral phosphate-binding aluminum gels for nondialyzed and dialyzed uremic patients as a method of controlling serum phosphorus concentrations [1]. It has recently been shown [2] that some aluminum is absorbed by uremic patients receiving these compounds. Since aluminum has a neurotoxic effect, it represents a potential health hazard when it is absorbed from orally administered salts [3-5].

The role of aluminum is not known in detail, but it has been shown that elevated aluminum concentrations are found in various areas of the brains of patients suffering from Alzheimer's presenile dementia [6].

The possible association between aluminum and dialysis dementia, a progressive encephalopathic syndrome in hemodialysis patients [7,8], has accordingly given rise to improved determinations of aluminum in serum [9] and other biological materials over the last two decades.

2. DETERMINATION OF THE TOTAL CONTENTS OF ALUMINUM[*]

2.1. Spectrophotometry

The spectrophotometric (or fluorimetric) determination of aluminum in clinical chemistry is no longer in use because the sensitivity is too low [10,11]. The last publications dealing with these methods are--according to the best knowledge of the authors--about 15 to 20 years old [12,11].

2.2. Atomic Absorption and ICP Atomic Emission Spectroscopy

A frequently applied method of measuring small amounts of aluminum in biological materials is AAS or AES. Aluminum at these low levels (0.1-1 mg/liter) has mostly been measured by AAS [13,14] (cf. Table 1). A serious problem, however, is that the actual matrix encountered in the analysis of biological materials is different from the ones reported earlier, and this fact led Krishnan et al. [15] to develop a method of analysis of biological materials for very small amounts of aluminum using AAS.

Krishnan et al. [15] studied the absorbance of aluminum with respect to fuel composition, slit width, wavelength, and the height

---

[*]1 mg/liter = 37 µmol/liter (Al).

of the light beam above the burner head in order to obtain the best sensitivity. Absorption measurements on a 50 mg/liter aluminum solution were carried out in the presence of various ions at the level of 5000 mg/liter and only five, Na^+, Zn^{2+}, La^{3+}, PO_4^{3-}, and Fe^{2+}, showed any significant effect. It was easy to compensate for Na^+, and further measurements revealed that the other four ions do not have any significant effect at concentrations below 1000 mg/liter. The aluminum blank could be reduced to 0.05 mg/liter. A detection limit of 0.1 mg/liter could be achieved in the solutions analyzed, and experiments using standards gave the values of ±10% for precision at the 1 mg/liter level.

Fuchs et al. [16] reported a method for the determination of aluminum in serum by flameless AAS. It was emphasized that only 25 µl serum was required and that sample preparation was not needed. It was pointed out that calibration curves were linear up to 0.1 mg/liter and that the detection limit was 1.1 µg/liter. The accuracy of the method was tested by deuterium background compensation, by recovery experiments, and by the dilution method. The precision of the method was tested and the coefficient of variation was found to be 5.3%. The concentration of aluminum in serum was found to be 0.037 mg/liter ranging from 0.01 to 0.09 mg/liter. Problems concerning contamination of aluminum from the different vessels were examined, and it was shown to be possible to avoid this kind of contamination by showing extreme care when performing the analyses.

LeGendre and Alfrey [17] published a flameless AAS method by means of which picogram amounts of aluminum in biological tissue can be determined. The authors took advantage of the high formation constant between the aluminum ion and the ethylenediamine-tetraacetate ion ($\approx 10^{16}$). The procedure consists of mixing samples of 10-25 mg of dry defatted bone, and muscle or ashed brain with 5 ml of a saturated solution of disodium ethylenediaminetetraacetate for 2-4 hr. The described extraction procedure combined with flameless AAS has been shown to minimize tissue handling, and sources of external aluminum contamination have virtually been avoided. It was shown

TABLE 1

Overview on the Methods Used for the Determination of Aluminum

Method	Detection limit	Sensitivity	Standard deviation	Pretreatment of sample and analytical range	Biological material	Ref.
Spectrophotometry						[10-12]
AAS				0.1-1 mg/liter		[13]
AAS	0.1 mg/liter			>2 mg		[14]
AAS			0.1 mg/liter			[15]
Flameless AAS	1.1 µg/liter			0.01-0.09 mg/liter	Serum	[16]
Flameless AAS	~(1-5) pg			Extraction with EDTA	Tissue	[17]
Flameless AAS				Dry ashing	Brain, tissue	[18]
Flameless AAS	0.05 mg/liter			Direct determ. 0.05-0.59 mg/liter	Whole blood	[19]
Flameless AAS				Investigation on possible reactions between Al and H, O, N, Cl, and S	Theoretical study	[20]
Flameless AAS				Wet digestion	Urine, bone	[21]

Method	Detection limit	Notes	Sample	Ref.
Flameless AAS	0.04 ng		Tissue	[22]
Flameless AAS	0.078 ng 0.03 ng		Serum, plasma urine	[23]
Flameless AAS	1.2 µg/liter 0.5 µg/liter	Dry ashing 33-35 µg/liter (serum) 19.3-20.3 µg/liter (cerebrospinal fluid)	Serum Cerebrospinal fluid	[25]
Flameless AAS		Investigation on methods for removal of chloride interferences	Theoretical study	[26]
Flameless AAS			Serum	[27]
ICP-AES	0.4 µg/liter (water) 1.0 µg/liter (urine) 4 µg/liter (blood)	Direct determinat. 2.2-7.2 µg/liter (urine) 8.5-16.5 µg/liter (blood)	Water, urine and blood	[28]
ICP-AES		Dry ashing	Blood, urine, feces, animal feed	[29]
Flameless AAS	2 µg/liter	Dry ashing 0.1-12.0 µg/liter	Serum	[9]

that various ions in the amount of 1 g/liter did not have any significant effect on the aluminum determination when added to a 100
µg/liter aluminum solution.

McDermott and Whitehill [18] also reported the determination
of aluminum in human and animal brain and in rabbit and human tissues
using flameless AAS. It was shown that sample preparation by dry
ashing was preferable for an acid digestion method (H_2SO_4, $HClO_4$, and
HNO_3 in Pyrex glassware at 160°C).

Langmyhr and Tsalev [19] performed AAS determinations of aluminum in heparinized and hemolyzed samples of undiluted whole human
blood. A direct method has been used whereby 2 µl of blood has been
pipetted into a graphite cup atomizer and after drying the aluminum
is determined at 2500°C. In an indirect method, 15 µl of blood is
decomposed by HNO_3 (90°C for 2 hr) and 2-µl portions of the solution
are then analyzed as in the direct method. The detection limit of
the direct method was 0.05 mg/liter aluminum, and the results ranged
from 0.05 to 0.59 mg/liter with a mean value of 0.20 mg/liter. The
direct analysis, the analysis by wet ashing, and direct analysis of
diluted whole blood (1 + 10) agreed well, and the differences between
the mean values have been explained by the presence of random errors
alone.

Persson et al. [20] investigated the reactions which might
occur between Al and H, O, N, Cl, and S when Al was determined by
flameless AAS. Their results have indicated that even in the 1-10
ng/liter range the presence of the elements mentioned have caused
severe interferences during the atomization step (2300-2900 K). It
has been shown that conditions for the formation of interfering compounds such as $AlO(g)$, $Al_2O(g)$, $Al_2O_3(s)$, $AlOH(g)$, $AlH(g)$, $AlHO_2(g)$,
$AlS(g)$, $AlN(g)$, $AlN(s)$, and $AlCl(g)$ exist. The influence on the
calculations of the equilibria for the reactions involving carbon-
oxygen and carbon-sulfur has been proved by varying the input amount
of carbon.

Garmestani et al. [21] used flameless AAS and described how
direct atomization of urine and bone dissolved in nitric acid resulted

in aluminum values that were systematically higher for bone and variable for urine, as compared with an NAA procedure. This was the result of calcium and phosphate interference of the aluminum signal. The optimum procedure involved digestion of bone or urine at 60°C with concentrated nitric acid and a solution containing $CaCl_2$ and K_2HPO_4 to negate the effects of varying calcium and phosphate contents in the samples.

Julshamn et al. [22] determined aluminum in human tissue samples using standard addition and flameless AAS, where the flameless method was shown to be superior to AAS using the nitrous oxide-acetylene flame. Aluminum levels down to about 0.1 mg/kg of freeze-dried tissue and 0.01 mg/liter serum can be determined after standard addition. The detection limits and sensitivities were 0.04 and 0.078 ng of aluminum, respectively.

Gorsky and Dietz [23] determined aluminum in biological fluids (serum, plasma, and urine) and have also used flameless AAS. The sensitivity was 0.03 ng, and the analytical recovery of aluminum added to 10 samples of serum with Al concentrations ranging from 4 to 136 µg/liter was 101.2 ± 7.2. The atomic absorption of the materials examined containing chloride was increased by adding sodium sulfate and thus changing the proportion of anion that is present as the chloride complex. It was found, though, not necessary in practice to add sulfate by analyzing serum and urine in the presence or absence of this ion (SO_4^{2-}), and the method of analysis is therefore less prone to contamination. The sensitivity is greater than that reported by Blotcky et al. [24] for the analysis of aluminum in urine by NAA.

Pegon [25] determined aluminum in serum and cerebrospinal fluid using flameless AAS. A wetting agent (Teepol) was added to shorten the time of drying and ashing steps. Aluminum was converted to aluminate with an aqueous concentrated ammonia solution to prevent volatile chloride formation. The mean value for the sera of 20 normal persons was 34.1 µg/liter, and the mean value for the cerebrospinal fluid was 19.8 µg/liter with relative standard deviations of 3.5 and 2.5%, respectively.

Matsusaki et al. [26] investigated methods for removal of
chloride interferences in determination of aluminum by flameless
AAS. Two mechanisms of chloride reactions have been examined. The
first arises from coordination of the chloride to aluminum. This
interference could be removed by preventing the coordination by controlling the ashing temperature at above 1000°C or by ashing ammonium
acetate or nitric acid. The other interference is due to coexisting
chloride salts remaining until the atomization step. This complica-
tion could be removed by volatilizing these coexisting chloride salts,
but not aluminum, or by converting them and/or aluminum chloride into
another substance, such as the oxides, before the atomization step.
The tetraamonium salt of EDTA has been shown to be very suitable as
an additive to overcome chloride interference because of its ability
to coordinate aluminum and to form ammonium chloride which is easily
volatilized.

King et al. [27] examined the binding of aluminum in the serum
of a normal male using gel filtration chromatography and flameless
AAS. The elution profile obtained with a gel filtration column separated the aluminum into four major peaks, which seemed to be associated with high-molecular-weight protein(s), albumin, possibly low-
molecular-weight protein(s) and/or some inorganic anions. The elution profile for a renal dialysis patient gave results which were
very similar. The aluminum concentration of serum was found to be
50 μg/liter of a normal male and 164 μg/liter of a patient with
chronic renal failure before routine hemodialysis was started.

Allain and Mauras [28] determined aluminum in water, urine,
and blood by inductively coupled plasma emission spectroscopy using
a concentric pneumatic nebulizer. Interferences were systematically
studied using different metals and metalloids commonly found in biological samples. Some metals, particularly Ca, Li, Sr, Na, and Fe,
increased background intensity, and alkali metals and alkaline earth
metals increased the net signal of Al, but both interferences could
be avoided by using the additive technique and measuring the background at peak base by wavelength shifts. The limits of detection
were 0.4 μg/liter in water, 1 μg/liter in urine, and 4 μg/liter in

blood. Aluminum assays on 14 healthy persons gave the following
results: blood 12.5 µg/liter and urine 4.7 µg/liter.

Lichte et al. [29] determined aluminum in blood, urine, feces,
and animal feed. The samples were ashed, fluxed with sodium carbonate,
and dissolved in HCl. The solutions were analyzed by inductively coupled plasma emission spectroscopy and the 394.4-nm emission line was
used. The method was tested by analyzing samples and pure water to
which known quantities of aluminum had been added. The precision of
the method was found to be rather limited by the instability of the
biological samples.

Alderman and Gitelman [9] described a method for determination
of aluminum in serum by flameless AAS. Interferences have been controlled and total combustion of the samples achieved by the use of a
new diluent in a molybdenum-treated graphite tube. Contamination
has been minimized by the use of carefully rinsed tubes and vessels.
The detection limit was found to be 2 µg/liter. The authors observed
a lower range of serum aluminum (0.1-12.0 µg/liter) than has been
reported elsewhere [23], and they assume that this finding has been
related to the absence of contamination in their specimens because
the recovery of aluminum added to normal sera is virtually complete.

ABBREVIATIONS AND DEFINITIONS

AAS	atomic absorption spectroscopy
AES	atomic emission spectroscopy
Detection limit	concentration that gives a signal-to-noise ratio of 2
EDTA	ethylenediaminetetraacetate
(g)	gas
ICP	inductively coupled plasma
NAA	neutron activation analysis
(s)	solid
Sensitivity	concentration that produces an absorbance of 0.0044 (1% absorption)

REFERENCES

1. M. E. Rubini, J. W. Coburn, and S. G. Massay, *Arch. Intern. Med.*, *124*, 663 (1969). Cit. from Ref. 17.

2. E. M. Clarkson, V. A. Luck, and W. V. Hynson, *Clin. Sci.*, *43*, 519 (1972).

3. L. Kopeloff, S. Barrera, and N. Kopeloff, *Am. J. Psychiatry*, *99*, 881 (1942).

4. I. Klatzo, H. Wisniewski, and E. Streicher, *J. Neuropathol. Exp. Neurol.*, *24*, 187 (1965). Cit. from Ref. 17.

5. C. A. Miller and F. M. Levine, *J. Neurochem.*, *22*, 751 (1974). Cit. from Ref. 17.

6. D. R. Crapper, S. S. Krishnan, and A. J. Dalton, *Science*, *180*, 511 (1973).

7. A. C. Alfrey, G. R. LeGendre, and W. D. Kaehny, *N. Engl. J. Med.*, *294*, 184 (1976). Cit. from Ref. 9.

8. J. R. McDermott, A. I. Smith, and M. K. Ward, *Lancet*, *1*, 901 (1978). Cit. from Ref. 9.

9. F. R. Alderman and H. J. Gitelman, *Clin. Chem.*, *26*(2), 258 (1980).

10. C. H. R. Gentry and L. G. Sherrington, *Analyst*, *71*, 432 (1946).

11. J. W. Poole, L. A. Zeigler, and M. A. Dugan, *J. Pharm. Sci.*, *54*, 651 (1965).

12. M. Seilbold, *Klin. Wochenschr.*, *38*, 117 (1960). Cit. from Ref. 19.

13. M. D. Amos and P. E. Thomas, *Anal. Chim. Acta*, *32*, 139 (1965).

14. D. Myers, *At. Absorp. Newslett.*, *6*, 89, July-Aug. (1967).

15. S. S. Krishnan, K. A. Gillespie, and D. R. Crapper, *Anal. Chem.*, *44*, 1469 (1972).

16. C. Fuchs, M. Brasche, K. Paschen, H. Nordbeck, and E. Quellhorst, *Clin. Chim. Acta*, *52*, 71 (1974).

17. G. R. LeGendre and A. C. Alfrey, *Clin. Chem.*, *22*(1), 53 (1976).

18. J. R. McDermott and I. Whitehill, *Anal. Chim. Acta*, *85*, 195 (1976).

19. F. J. Langmyhr and D. L. Tsalev, *Anal. Chim. Acta*, *92*, 79 (1977).

20. J.-Å. Persson, W. Frech, and A. Cedergren, *Anal. Chim. Acta*, *92*, 85 (1977).

21. K. Garmestani, A. J. Blotcky, and E. P. Rack, *Anal. Chem.*, *50*, 144 (1978).

22. K. Julshamn, K.-J. Andersen, Y. Willassen, and O. R. Braekkan, *Anal. Biochem.*, *88,* 552 (1978).

23. J. E. Gorsky and A. A. Dietz, *Clin. Chem.*, *24*(9), 1485 (1978).

24. A. J. Blotcky, D. Hobson, J. A. Leffler, E. P. Rack and R. R. Recker, *Anal. Chem.*, *48,* 1084 (1976).

25. Y. Pegon, *Anal. Chim. Acta*, *101,* 385 (1978).

26. K. Matsusaki, T. Yoshino, and Y. Yamamoto, *Talanta,* *26,* 377 (1979).

27. S. W. King, M. R: Wills, and J. Savory, *Res. Commun. Chem. Pathol. Pharmacol.*, *26,* 161 (1979).

28. P. Allain and Y. Mauras, *Anal. Chem.*, *51,* 2089 (1979).

29. F. E. Lichte, S. Hopper, and T. W. Osborn, *Anal. Chem.*, *52,* 120 (1980).

Chapter 13

DETERMINATION OF GOLD IN CLINICAL CHEMISTRY

Arne Jensen
Chemistry Department AD
Royal Danish School of Pharmacy
Copenhagen, Denmark

Erik Riber and Poul Persson
Medi-Lab a.s.
Copenhagen, Denmark

and

Kaj Heydorn
Isotope Division
Risø National Laboratory
Roskilde, Denmark

1. CLINICAL CHEMICAL ASPECTS

The therapeutic use of gold compounds is of considerable antiquity, as discussed by Nineham [1]; it has been used for the treatment of patients undergoing chrysotherapy for rheumatoid arthritis since

1927[2]. The treatment has over the years been largely empirical until the estimation of gold could be performed by means of NAA or AAS. Until the advent of the methods mentioned, the estimation of this element in biological fluids by classical chemical or other methods was beyond the capability of the clinical laboratory, and the therapeutic effect of gold was statistically first proved in the 1960s.

The treatment for rheumatoid arthritis consists in general of injection of a dissolved gold(I) complex; since the accumulation of high concentrations of gold in the bloodstream has undesirable effects, e.g., allergic side effects as skin exanthemes, the monitoring of gold in the serum and urine of patients undergoing such treatment is important.

On the other hand, gold has not yet been identified in humans as a prerequisite for normal life processes and at present the gold concentration in human blood before gold treatment is below detectable levels (see Sec. 3). During treatment the level in blood serum, the blood fraction that has been most extensively studied, is between 1 and 10 mg/liter [3,4].

2. DETERMINATION OF THE TOTAL CONTENTS OF GOLD[*]

2.1. Atomic Absorption Spectroscopy

A number of authors have reported the estimation of gold in serum by AAS. Few details of the method of Frajola and Mitchell [2] are mentioned in the paper by Dunckley [7]. Lorber et al. [5] use an internal standard method which, while it satisfactorily compensates for matrix effects, requires relatively large amounts of serum. Dietz and Rubinstein used a dilution of serum (1:1) and experienced pronounced matrix effects, which could be attributed to the presence of protein [6].

---

[*]1 mg/liter = 5.1 μmol/1 (Au)

Dunckley [7] reported an AAS method for the estimation of gold in diluted serum. The flame was slightly fuel-rich and the wavelength used was 242.8 nm. Matrix effects are compensated by adding protein to the working standards, and the detection limit was 50 µg/liter. Sera and standards were extracted into methyl isobutyl ketone for AAS. The interbatch variance was 3.2%.

Balázs et al. [8] described a method for the determination of total gold levels by AAS using 242.8 nm with emphasis on plasma and urine gold. Gold was administered as sodium gold(I) thiomalate. However, the direct extraction of trivalent gold halide anion complexes from strong acid into organic solvents was shown to result in a very clean extract which was both rapid and complete. The authors used acid permanganate for simultaneously releasing gold from its very strong protein binding and oxidizing it to the aurate(III) form. 6 M HCl was used converting gold to hydrogen tetrachloroaurate(III) which was extracted into methyl isobutyl ketone. A reliable detection limit for gold was 10 µg/liter (urine) and 50 µg/liter (serum).

Aggett [9] established that the determination of gold in serum by AAS can be carried out with higher sensitivity with a carbon filament for atomization, i.e., flameless AAS. A comparison between flame atomization and carbon filament atomization showed good agreement between the two sets of data, and the flameless AAS method gave a relative standard deviation of 4%, as determined from 15 measurements on one serum sample.

Dunckley [10] described an AAS method for estimating gold in the urine of patients undergoing chelation therapy, but there is no indication of which chelate of gold has been used. Variable absorption at 242.8 nm due to background has been avoided by wet ashing and extraction of gold into methyl isobutyl ketone. Replicate analysis of a chelation therapy urine (24 determinations) with a mean concentration of 0.64 mg/liter showed a variance of 8.3%. The detection limit was shown to be 10 mg/liter.

Maessen et al. [11] performed a direct determination of gold (cobalt and lithium) in blood plasma at the mg/liter level by means

of flameless AAS using the Mini-Massmann carbon rod atomizer and a
suspension of gold(I) thioglucose in vegetable oil mixed with living
plasma. The authors have experimentally examined sample introduction
and alteration of the properties of graphite. It was shown that a
very pronounced gain in precision was obtained both by means of in-
strumentally controlled sample introduction and by coating of plunger
and needle of the syringe with polystyrene thus preventing the adsorption of metal ions on the syringe. The authors compared the analytical
results for gold obtained with both the carbon rod atomizer (as used
in AAS) and a gas-stabilized direct-current arc (as used in atomic
emission spectroscopy), and no systematic differences were apparent.
A thousand times smaller sample volume per analysis could, however,
be used in the case of the carbon rod.

Kamel et al. [12] compared procedures that involve the combination of the use of either carbon rod or carbon furnace atomization
and AAS with the purpose of determining the level of gold in whole
blood, plasma, and serum from patients undergoing gold treatment.
It has been shown that the carbon furnace atomization was the preferred procedure because of the high sensitivity. The detection
limits for gold were, with the preferred procedure and administered
as sodium gold(I) tiomalate, 2, 2, and 4.5 μg/liter in serum, plasma,
and whole blood, respectively. The method has been used to confirm
that most gold is carried in the serum fraction of the blood and that
the concentration of gold in human blood before treatment is below
detectable levels. During treatment the level was found to be 1-10
mg/liter.

Schattenkirchner and Grobenski [13] used the combination of
AAS and either graphite furnace and auto sampler or flame technique,
respectively, for measurements in blood and urine during gold treatment. Background corrections and the method of additions were used,
and no difference between the gold content in serum and blood was
found. The gold was administered as gold(I) thiopolypeptide, and
the concentration of the element in plasma and urine was followed
during the first 42 days of treatment with this complex.

Ward et al. [14] analyzed gold using flame and electrothermal AAS and NAA (cf. Sec. 3). The three techniques were compared using plasma and plasma fractions (derived by gel chromatography) from rheumatoid patients receiving gold(I) thiomalate therapy and from plasma samples to which were added gold(I) thiomalate in vitro. The three methods agreed well in the analysis of gold in whole plasma, but only NAA could be used for the assay of all the plasma fractions. Sodium chloride had no effect on the gold signal with the flame technique, but as the concentration of this salt increased, there was a steady decrease in the gold atomic signal when using electrothermal AAS. The reason for the lower results yielded by this method when compared with the two other techniques could not be explained.

Barrett et al. [15] used an assay for gold in whole blood of arthritis patients applying the graphite furnace AAS procedure. No pretreatment of the whole blood was necessary except for simple dilution, thereby eliminating some variables and saving time. The assay developed is very useful because the chrysotherapy can be stopped if a patient shows no clinical improvement even at the therapeutic blood levels (3-6 mg/liter).

Wawschinek and Rainer [16] stated that gold(I) has been extracted from serum and urine by means of dimorpholinethiuram disulfide dissolved in methyl isobutyl ketone as reagent-extractant solution. Thus, the usual oxidation of gold(I) to gold(III) before extraction was avoided. It was, however, necessary to use this specific reagent-extractant in order to obtain 97-99% recovery since only 6-8% of the serum or urine gold was recovered when the ketone mentioned was used without the chelating reagent. The content of gold was then determined using flameless AAS and the matrix interferences were avoided by using the extraction procedure. Gold standards were prepared from sodium gold(I) thiomalate (Tauredon) because this compound is the usual one used for chrysotherapy. In this connection it might be added that a few other gold(I) complexes, e.g., sodium gold(I) thiosulfate, have been used for therapeutic purposes [2].

Wawschinek [17] also suggested another procedure for determination of gold (and copper) in serum by flameless AAS. A dilution

mixture of phosphoric acid, ammonium nitrate, and a tenside (Triton
X-100) is used for serum samples. This mixture allows higher ashing
temperature and avoids matrix disturbance. The standards were prepared from sodium gold(I) thiomalate (Tauredon) and a recovery of
98-99% was obtained.

Melethil et al. [18] developed a flameless AAS assay capable
of accurately determining nanogram amounts of gold bound to 2 and 4%
bovine serum albumin in 0.1 M phosphate buffer. It was shown that
the overall relative binding values of gold were 98 and 99%, respectively. Interaction studies were also carried out which revealed
that gold was not displaced from the binding sites by salicylic acid.

2.2. Spectrophotometry

Spectrophotometry should also be included among analytical methods
which have been used for determination of gold in biological fluids
[19]. However, this method requires extensive sample manipulation,
is time consuming, and suffers from low sensitivity. These disadvantages make colorimetry (or spectrophotometry) inapplicable to the
routine determination of therapeutic gold concentrations (1-10 mg/
liter). This optical method is therefore outdated as an analytical
tool for determination of this element in clinical chemical analysis.

3. NEUTRON ACTIVATION ANALYSIS

The determination of gold in biological material by neutron activation analysis at the normal level of concentration was attempted in
1949 by Tobias and Dunn [20] in their first demonstration of neutron
activation analysis with radiochemical separation for the analysis
of biomedical samples.

The very high sensitivity for gold is based on the use of the
indicator isotope [198]Au, which has a half-life of 2.7 days and a
characteristic energy of 412 keV. Even then, the extremely low

concentrations in serum and most other tissues from patients without ingestion of gold-containing drugs cannot be determined with satisfactory precision by instrumental neutron activation analysis (cf. Chap. 6). Decomposition of the sample after irradiation followed by a complete radiochemical separation is required to eliminate the interference from other elements.

Early attempts to determine gold in normal human serum were made by Sølvsten [21] and by Bagdavadze and co-workers [22], who found an average concentration of 60 ng/liter in excellent agreement with more recent data [23,24].

Parr and Taylor [25] determined concentrations of gold in normal human livers and they found very considerable variation among 32 samples analyzed. The observed mean value of approximately 0.1 µg/kg wet weight is in substantial agreement with modern results for liver [26,27].

The insufficient sensitivity of other analytical methods prevents confirmation of these results by independent analysis, but at the much higher concentrations encountered in patients undergoing chrysotherapy such comparisons have been made. At these levels, however, gold may be determined by instrumental neutron activation analysis without radiochemical separation. Thus Ward et al. [14] found satisfactory agreement between atomic absorption spectrophotometry and neutron activation analysis for plasma concentrations of 0.5-20 mg/liter, while Turkall and Bianchine [28] found slightly higher results by NAA for tissue samples at the approximate level of 1 mg/kg wet weight.

The conclusion is that both AAS and NAA may be used for the monitoring of patients undergoing chrysotherapy whereas only neutron activation analysis with radiochemical separation is applicable to the measurement of normal levels of gold in biomedical samples.

An overview of the methods used for gold determination is given in Table 1.

TABLE 1

Summary of Methods Used for the Determination of Gold

Method	Detection limit (μg/liter)	Pretreatment of sample and analytical range	Biological material	Ref.
Flameless AAS		1-10 mg/liter	Serum	[3,4]
AAS		--	Serum	[2]
AAS		--	Serum	[5]
AAS		1:1 dilution	Serum	[6]
AAS	50	Extraction + dilution	Serum	[7]
AAS	10 (urine) 50 (serum)	Extraction of H_3O^+, $AuCl_4^-$	Urine Serum	[8]
AAS Flameless AAS			Serum	[9]
AAS	10	Wet ashing + extraction 0.6-0.7 mg/liter	Urine	[10]
Flameless AAS		Direct determ. ~1 mg/liter	Plasma	[11]
Flameless AAS	2 2 4.5	-- 1-10 mg/liter	Serum Plasma Whole blood	[12]
AAS Flameless AAS			Urine Plasma	[13]
AAS Flameless AAS			Plasma	[14]
NAA			Plasma fractions	[14]
Flameless AAS		Dilution 3-6 mg/liter	Whole blood	[15]
Flameless AAS		Extraction of Au(I)	Serum Urine	[16]
Flameless AAS		Wet ashing after dilution with H_3PO_4, NH_4NO_3, and Triton X-100	Serum	[17]
Flameless AAS			Serum	[18]
Spectrophotometry				[19]
NAA		No pretreatment	Blood	
NAA		No pretreatment	Tissue	

ABBREVIATIONS AND DEFINITIONS

AAS	atomic absorption spectroscopy
Detection limit	concentration that gives a signal-to-noise ratio of 2
NAA	neutron activation analysis
Sensitivity	concentration that produces an absorbance of 0.0044 (1% absorption

REFERENCES

1. A. R. Nineham, *Arq. Interamer. Reumatol.*, *6*, 113 (1963). Cit. from Ref. 7. See also K. C. Dash and H. Schmidbaur, Gold complexes as metallo-drugs, in *Metal Ions in Biological Systems*, Vol. 14 (H. Sigel, ed.), Marcel Dekker, New York, 1982, p. 179.

2. W. J. Frajola and P. B. Mitchell, *Fed. Proc.*, *26*, 780 (1967). Cit. from Ref. 7.

3. J. D. Jessop and R. G. S. Johns, *Ann. Rheum. Dis.*, *32*, 228 (1973). Cit. from Ref. 12.

4. A. Lorber, C. J. Atkins, C. C. Chang, Y. B. Lee, J. Starrs, and R. A. Bovy, *Ann. Rheum. Dis.*, *32*, 133 (1973). Cit. from Ref. 12.

5. A. Lorber, R. J. Cohen, C. C. Chang, and H. E. Anderson, *Arthr. Rheum.*, *11*, 170 (1968). Cit. from Ref. 7.

6. A. A. Dietz and H. M. Rubenstein, *Clin. Chem.*, *15*, 787 (1969) (Abstr.). Cit. from Ref. 7.

7. J. V. Dunckley, *Clin. Chem.*, *17*, 992 (1971).

8. N. D. H. Bálazs, D. J. Pole, and J. R. Masarei, *Clin. Chim. Acta*, *40*, 213 (1972).

9. J. Aggett, *Anal. Chim. Acta*, *63*, 473 (1973).

10. J. V. Dunckley, *Clin. Chem.*, *19*(9), 1081 (1973).

11. F. J. M. J. Maessen, F. D. Posma, and J. Balke, *Anal. Chem.*, *46*, 1445 (1974).

12. H. Kamel, D. H. Brown, J. M. Ottoway, and W. E. Smith, *Analyst*, *101*, 790 (1976).

13. M. Schattenkirchner and Z. Grobenski, *At. Absorp. Newslett.*, *16*, 84 (1977).

14. R. J. Ward, C. J. Danpure, and D. A. Fyfe, *Clin. Chim. Acta*, *81*, 87 (1977).

15. M. J. Barrett, R. DeFries, and W. M. Henderson, *J. Pharm. Sci.*, *67*, 1332 (1978).

16.   O. Wawschinek and F. Rainer, *Atom. Absorp. Newslett.*, *18*, 50 (1979).

17.   O. Wawschinek, *Mikrochim. Acta, 1979*(II), 111.

18.   S. Melethil, A. Poklis, and V. A. Sagar, *J. Pharm. Sci.*, *69*, 585 (1980).

19.   J. F. Goodwin and A. J. Bollet, *J. Lab. Clin. Med.*, *55*, 965 (1960). Cit. from Ref. 18.

20.   C. A. Tobias and R. W. Dunn, *Science*, *109*, 109 (1949).

21.   S. Sølvsten, *Scand. J. Clin. Lab. Invest.*, *16*, 39 (1964).

22.   N. B. Bagdavadze, L. V. Barbakadze, E. N. Ginturi, N. E. Kuchava, A. M. Mosulishvili, and N. E. Kharabadze, *Soobsheh Akad. Nauk. Gruz. SSR, 39,*287 (1965).

23.   N. I. Ward and D. E. Ryan, *Anal. Chim. Acta, 105*, 185 (1979).

24.   K. Kasperek, G. V. Iyengar, J. Kiem, H. Borberg, and L. E. Feinendegen, *Clin. Chem.*, *25*, 711 (1979).

25.   R. M. Parr and D. M. Taylor, *Phys. Med. Biol.*, *8*, 43 (1963).

26.   D. Brune, G. F. Nordberg, P. O. Wester, and B. Bivered, in *Nuclear Activation Techniques in the Life Sciences 1978* (Proceedings of a Symposium), International Atomic Energy Agency, Vienna, 1979, p. 643ff.

27.   L.-O. Plantin, Ref. 26, p. 321ff.

28.   R. M. Turkall and J. R. Bianchine, *Analyst, 106*, 1096 (1981).

Chapter 14

DETERMINATION OF PHOSPHATES IN CLINICAL CHEMISTRY

Arne Jensen
Chemistry Department AD
Royal Danish School of Pharmacy
Copenhagen, Denmark

and

Adam Uldall
Department of Clinical Chemistry
University of Copenhagen, Herlev Hospital
Herlev, Denmark

1. CLINICAL CHEMICAL ASPECTS

1.1. Physiological and Pathophysiological Chemistry

Phosphate is involved in many biological processes, e.g., as glucose-6-phosphate in the Krebs cycle and as energy-rich phosphate in adenosine triphosphate [1,2].

The total amount of phosphates in the adult human body is approximately 20 mol; nearly 80% of this amount is incorporated in the skeleton, mainly as hydroxyapatite, approx. $Ca_5(PO_4)_3(OH)$. It crystallizes in thin plates with a total surface of 10,000-20,000 m^2 and serves here as a buffer for calcium ion and phosphate in the intercellular fluid. The concentration of phosphate and phosphate-containing components is much higher in the intracellular liquid than in the extracellular fluid (approximately 50 times higher).

The daily supply of phosphate-containing compounds in food is in the magnitude of 1/20 mol and is normally sufficient. Alkaline phosphatases in the intestine fluid hydrolyze phosphate esters and 2/3 of the free phosphate is normally absorbed. Phosphate is eliminated through the kidney by incomplete reabsorption after glomerular filtration.

As a general rule the regulation of phosphate is passive and secondary to, say, acid/base balance, calcium metabolism, and glycolysis. Parathyroid hormone, however, increases the urinary output of phosphate due to inhibition of the tubular reabsorption. At the same time the hormone increases the dissolution of calcium phosphate from the skeleton and more phosphate will therefore pass into the intracellular fluid. Simple solubility conditions imply that the calcium phosphate in the skeleton will dissolve in uncompensated acidosis and increase the concentrations of calcium(II) and phosphate in the intercellular fluid; the increased concentration of calcium ion will deliver calcitonin from thyroidea and promote precipitation of calcium phosphate in the skeleton. Precipitation of calcium phosphate takes place in uncompensated alkalosis and the low calcium concentration promotes delivering of parathyroid hormone, which increases

the calcium concentration by increasing the tubulary reabsorption
and by dissolving skeleton calcium phosphate. The effect of vitamin
D on the calcium metabolism will also be reflected in the phosphate
metabolism. One-third of the total phosphates in plasma (and serum)
is nonester- or amide-bound (inorganic) and the major residual parts
are phospholipids (lecitins). Phospholipids in lipoproteins are
essential for their capability to keep triglyceride and cholesterol
in solution in the body fluids. The excretion of calcium ion and
phosphate through the kidney involve the possibility of precipitation
of calcium phosphates in the urinary tract. Inhibitors such as
citrate, magnesium, pyrophosphate, and mucopolysaccharides normally
prevent formation of more types of urinary tract calculi (cf. Chap. 4).

1.2. Hospital Laboratory Aspects

Results of estimation of phosphate in fluids of patients are not
used in the early diagnosis, and the determinations are generally
not stat work. The most frequent estimations are the determination
of so-called inorganic phosphate ($\sim$ phosphate; cf. Sec. 1.3) in serum
which also includes a minor amount of phosphate delivered from, e.g.,
adenosine triphosphate. Furthermore, the determination of phosphate
in urine is carried out quite frequently; however, both analyses,
considered as well collectively and individually, are normally not
more frequent than a few percent of the total laboratory workload.
The much higher concentration of phosphate compounds inside the cells
than outside makes the handling of blood specimens critical, and a
quick and complete separation of serum from erytrocytes, leukocytes,
and thrombocytes is essential.

The analytical accuracy and precision is, at present, unfortunately not as high as could be desired; the overall variation is
estimated to 6.5% (relative standard deviation) [3]. One major
reason for this variation is probably minor differences between the
methods in use, and it is expected that development of an extremely

reliable reference method, a so-called definitive method, will facilitate a decrease of the interlaboratory variation [4].

Many other phosphate-containing compounds are measured in specialized laboratories, e.g., phospholipids in serum and cyclic adenosine-3',5'-monophosphate (cAMP) in blood or urine. Pyrophosphate is also of interest [5].

1.3. Nomenclature of Phosphates in Clinical Chemistry

The oxidation state of the elements have often been ignored in the nomenclature of biochemical literature. The designation *phosphorous* has been used for phosphates in living systems in spite of the fact that elemental phosphorous is a strong poison [2]. The term "phosphorous" has probably been preferred because no ambiguity is present when a value for a measurement of mixtures of different phosphates is stated as mass. The International Federation of Clinical Chemistry and the Clinical Chemistry Section of International Union of Pure and Applied Chemistry have recommended [6] referring to the amount of substance (mole) instead of the mass when reporting a result. In this situation the designation "phosphorous" should be used only for the oxidation step zero. The oxidation step of phosphorous in biological systems is normally five and the term "phosphate" is recommended [6]. If the reported result refers to more than one phosphate-containing component (covalent binding), the designation "phosphates" is used, and eventually an explanatory suffix is stated. The International Federation of Clinical Chemistry has recommended the term "phosphate" in the definition of inorganic phosphate [6]. This designation is used in the present work despite the fact that most of the routine methods for phosphate in serum also estimate a minor part of the organic-bound phosphate. The designation "phosphates" is used for unspecified combinations of unbound and/or covalent-bound phosphate compounds. The term "phosphates(total)" is used for the sum of all phosphate-containing components.

2. DETERMINATION OF PHOSPHATES WITH AID OF MOLYBDENUM

2.1. Determination of Phosphate Using Molybdenum Blue

Under suitable conditions molybdates react with phosphate to form
heteropoly compounds such as $(NH_4)_3[P(Mo_3O_{10})_4]$ [6]. The technique
for the determination of phosphate involves the *photometric determination of the molybdenum blue*, which is formed by reduction of
$(NH_4)_3[P(Mo_3O_{10})_4]$, but the mechanism is complex and the reaction
product is not known in detail [7]. The different reduction techniques used differ mainly in the *choice of reductor*:

1. Tin(II) chloride produces the most molybdenum blue from a
    given amount of the heteropoly compound, but this tin(II)
    salt as reductor gives an unstable color deviation from
    Beer's law and, because of instability of the reductor
    itself, difficulty in reproduction of standards from day
    to day. These difficulties can be avoided by addition of
    hydrazine [8].

2. Phenylhydrazine [9], hydroquinone [10], hexacyanoferrate(II)
    [11], and hydrazine sulfate [12] have, among other reductors,
    all been used.

3. Ascorbic acid can be used at pH 4 where the splitting of
    labile phosphate esters is 5% or less than that occurring
    with some of the other reductors mentioned [13].

4. Iron(II) sulfate has the advantage of color stability and
    greater specificity when labile phosphate esters are present,
    and the main reason is believed to be that the reaction is
    carried out in the presence of less acidity than that which
    occurs in most of the other methods [14].

Complex chemical considerations must be taken into account when the
analysis is performed on whole blood because a chance of splitting
organic esters in the course of the analysis may occur. A solution
to this problem is to add a citrate-arsenite buffer system which
forms complexes with the excess color reagent remaining after the
phosphate initially present has reacted. Then if organic phosphates

hydrolyze to release inorganic phosphate, there is no chromogen available to react with it [15].

2.2. Methods for Phosphate-Measuring Phosphomolybdate

The basis of these methods is that under well-defined conditions molybdates react with phosphate to form $(NH_4)_3[P(Mo_3O_{10})_4]$. The ammonium phosphomolybdate may then be determined by an acidimetric titration [16] or the absorbance of the unreduced species is measured at 340 nm, or after extraction with a mixture of xylene-isobutanol at 310 nm [17]. An indirect determination of phosphate has been carried out by extracting the phosphomolybdic acid using 2-octanol and then determining Mo by atomic absorption spectroscopy [18].

2.3. Determination of Total Phosphates

The total phosphate of blood is determined as nonester-bound phosphate after wet ashing with sulfuric acid and perchloric acid, nitric acid, hydrochloric acid, or hydrogen peroxide [19-25]. The phosphate is determined as indicated above (cf. page 239), and the addition of a citrate-arsenite buffer system avoids having a chromogen present to react with organic phosphate which may have been formed by hydrolysis (cf. page 239). The organic phosphate can, therefore, be determined from the amount of total phosphates less the amount of inorganic phosphorous. The inorganic phosphate cannot, in general, be determined in serum without removal of proteins, i.e., an iron(II)-trichloroacetic acid reagent is added and the supernate is mixed with molybdic acid, etc., as indicated above. It is, however, possible to use a direct method not requiring protein removal when ascorbic acid is used as reductor in presence of borate [26]. Many other approaches have been published; see, for example, Ref. 27.

3. OTHER METHODS OF ANALYSIS FOR
DETERMINATION OF PHOSPHATES

The formation of $H_3[P(Mo_3O_{10})_4]$ in strongly sulfuric acidic solution is established within a few seconds. This complex is reduced by ascorbic acid (cf. page 239). The rate of the reduction reaction can easily be controlled by the acidity, and in sulfuric acid the rate of formation of molybdenum blue as indicated by the change in absorbance at 650 nm with respect to time has been shown to be [28]:

$$\frac{dA}{dt} = \frac{k_1[PO_4^{3-}][Mo(VI)]^6[\text{ascorbic acid}]}{k_2[H^+]^{10} + k_3[\text{ascorbic acid}]}$$

where $\frac{dA}{dt}$ corresponds to the rate of the reaction. If the starting concentration of sulfuric acid, Mo(VI), and ascorbic acid are the same for each sample and somewhat in excess of the stoichiometric amount needed for the amount of phosphate present, the initial rate of absorbance change is directly proportional to the phosphate concentration, i.e., $\frac{dA}{dt_i} = k'[PO_4^{3-}]$. Thus only the initial portion of the curve is of immediate use for the kinetic determination of phosphate in this complex mechanism.

Another type of phosphate determination involves the formation of molybdivanadophosphoric acid. When an excess of molybdate is added to an acidified solution of an orthophosphate and a vanadate, a heteropolycomplex believed to have the formula $(NH_4)_3PO_4 \cdot NH_4VO_3 \cdot 16MoO_3$ is formed [7], and the stable yellow color of this complex is measured photometrically. This method has been used for the determination of phosphate in blood and serum [29].

A dye-binding method may also be used for the phosphate determination, which in this case takes advantage of the reaction of phosphomolybdate with the triphenylmethane dye malachite green, thus giving rise to a marked shift in the absorption maximum from that of the dye alone to that of the complex [30]. The method has the drawback that pH has to be around 0, in which case organic phosphates are hydrolyzed unless the temperature is kept at 5°C.

A phosphate redox electrode system has recently been utilized
for determination of inorganic phosphate in serum [31]. The electrode consisted of a chemically treated iron wire in a turbulent
flow-through cell at a constant oxygen tension.

Finally, it must be mentioned that certain reactions based on
enzymes and involving phosphate take place at neutrality, thereby
permitting measurement of phosphate in the presence of unstable
ester-bound phosphates. In the successive enzyme reactions NADPH
is at last produced from NADP in proportion to the phosphate present
and is measured by its fluorescence or light absorption [32].

REFERENCES

1. H. A. Harper, in *Review of Physiological Chemistry* (H. A. Harper,
    ed.), Lange, California, 1973, p. 419.

2. N. Weissman, in *Clinical Chemistry: Principles and Technics*
    (J. R. Henry, D. C. Cannon, and J. W. Winkelman, eds.), Harper &
    Row, New York, 1974, p. 720.

3. T. Aronson, P. Bjørnstad, E. Leskinen, A. Uldall, and C.-H.
    de Verdier, in *Assessing Quality Requirements in Clinical Chem-
    istry* (M. Hørder, ed.), Nordic Clinical Chemistry Project
    (NORDKEM), Helsinki, 1980, p. 11.

4. N. W. Tietz, *Clin. Chem.*, *25*, 833 (1979).

5. J.-B. Roulett, B. Lacour, A. Ulmann, and M. Bailly, *Clin. Chem.*,
    28, 134 (1982).

6. S. T. Woods and M. G. Mellon, *Anal. Chem.*, *13*, 760 (1941).

7. R. P. A. Sims, *Analyst*, *86*, 584 (1961).

8. M. Kraml, *Clin. Chim. Acta, 13*, 442 (1966).

9. A. E. Taylor and C. W. Miller, *J. Biol. Chem.*, *18*, 215 (1914).

10. J. H. Roe, O. J. Irish, and J. I. Boyd, *J. Biol. Chem.*, *67*, 579
    (1926).

11. F. F. Tisdall, *Arch. Pediatr.*, *39*, 559 (1922).

12. J. F. Goodwin, R. Thibert, D. McCann, and A. J. Boyle, *Anal.
    Chem.*, *30*, 1097 (1958).

13. M. H. Aprison and K. M. Hanson, *Proc. Soc. Exp. Biol. Med.*, *100*,
    643 (1959).

14. M. Rockstein and P. W. Herron, *Anal. Chem.*, *23*, 1500 (1951).

15. E. S. Baginski, P. P. Foa, and B. Zak, *Clin. Chim. Acta, 15,* 155 (1967).

16. R. Linder and P. L. Kirk, *Mikrochemie, 22,* 300 (1937).

17. R. H. Dreisbach, *Anal. Biochem., 10,* 169 (1965).

18. W. S. Zaugg and R. J. Knox, *Anal. Chem., 38,* 1759 (1966).

19. E. J. Baumann, *J. Biol. Chem., 59,* 667 (1924).

20. H. Behrendt, *Am. J. Dis. Child., 64,* 55 (1942).

21. E. H. Fiske and Y. Sulbarow, *J. Biol. Chem., 66,* 375 (1925).

22. C. Nehe, *Acta Med. Scand., 125,* 505 (1946).

23. S. E. Kerr and L. Daoud, *J. Biol. Chem., 109,* 301 (1935).

24. J. S. King, Jr. and R. Buchman, *Clin. Chem., 15,* 31 (1969).

25. J. H. Roe, O. J. Irish, and J. I. Boyd, *J. Biol. Chem., 67,* 579 (1926).

26. J. F. Goodwin, *Clin. Chem., 16,* 776 (1970).

27. J. M. Piper and S. J. Lovell, *Anal. Biochem., 117,* 70 (1981).

28. G. G. Guilbault, *Modern Quantitative Analysis Experiments for Non-Chemistry Majors,* Marcel Dekker, New York, 1974, p. 96.

29. S. G. Simonsen, M. Westman, L. M. Westover, and J. W. Mehl, *J. Biol. Chem., 166,* 747 (1946).

30. K. Itaya and M. Ui, *Clin. Chim. Acta, 14,* 361 (1966).

31. J. G. Montalvo, Jr., L. A. Truxillo, R. A. Wawro, T. A. Watkin, A. Phillips, and R. M. Jenevin, Jr., *Clin. Chem., 28,* 655 (1982).

32. D. W. Schultz, J. V. Personneau, and O. H. Lowry, *Anal. Biochem., 19,* 300 (1967).

Chapter 15

IDENTIFICATION AND QUANTIFICATION OF SOME DRUGS
IN BODY FLUIDS BY METAL CHELATE FORMATION

R. Bourdon, M. Galliot, and J. Hoffelt
Laboratory of Biochemistry, Hospital Fernand Widal
University of Paris V
Paris, France

1. INTRODUCTION

Metal chelate formation is a useful chemical property for identification of organic compounds, especially by spectroscopic methods.
Generally, such a reaction is quite specific and the chelate is
easily extracted by organic solvents, particularly the nonpolar,
non-water-miscible ones. Our purpose has been to establish analytical methods on such properties in order to quantify drugs in body
fluids. There are two cases:

245

1. Some drugs are able to give directly metal chelates, such
    as the barbiturates whose identification was performed many
    years ago by the Parri reaction (cobaltoammine complex) or
    by mercury chelate formation [1]. The quantification itself
    can be made at a microscale by colorimetry of the mercury
    chelate [2].

2. For other drugs, which have a priori no chelatable group,
    a simple quantitative chemical reaction may lead to a che-
    lating agent, e.g., Δ9-tetrahydrocannabinol, isoniazid, and
    colchicine.

The chelation reaction can either be performed alone or in connection with a separation method. The quantification itself is generally based on the absorption or the fluorescence of the chelates.
Atomic absorption spectrometry can also be used for an indirect
measurement of the chelate through its metal ion.

2. DIRECT CHELATION METHODS

2.1. Increase of Solubility in Nonpolar Solvents

This method is demonstrated with guanoxabenz (= 1-(2,6-dichlorobenzal-
aminol-3-hydroxyguanidinium chloride), which is a central α-mimetic
drug used as a hypotensive (Fig. 1). This compound is highly photo-
sensitive, thermolabile, and soluble in water, but its structure is
pH-dependent. This molecule is stable at $2 \leq pH \leq 6$, dehydroxylated
and cyclised at lower pH and dimerised at pH >6. Obviously, the
classical extraction method for amines cannot be used in the present
case, but the linkage $=N-NH-C\begin{smallmatrix}NH\\NHOH\end{smallmatrix}$ may indicate the possibility for
the formation of chelates with metal ions. Actually, guanoxabenz
forms chelates with Cu^{2+}, Hg^{2+}, Sc^{3+}, V^{5+}, and Au^{3+}. The copper
chelate, which is stable at $5.5 \leq pH \leq 6.5$, is easily soluble in
organic solvents; this allows a complete extraction from an aqueous
phase as long as the Cu^{2+} concentration is high enough and the buffer
used reacts only as a weak copper-complexing agent, e.g., cacodylate.

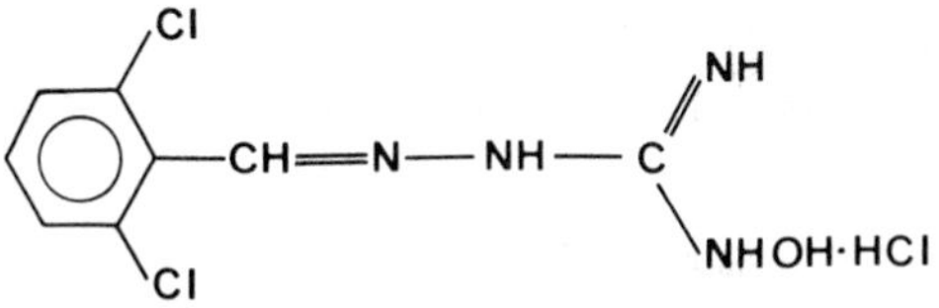

FIG. 1.   Guanoxabenz.

With the classical methods [3-5] the formation of a 1:1 complex can be demonstrated (K_1 = 2.5 x 10^5 M^{-1} at 20°C).

The copper chelate extraction can be performed directly with ethylacetate from body fluids; however, the recovery is better after absorption of the complex on Extrelut (Merck) and elution.   The copper present in the extract does not limit the other analytical procedures (hydrolysis and gas chromatography with electron capture detection). The technical details have been published elsewhere [6].

2.2.   Increase of Solubility in Water

Increase of water solubility is exemplified with cimetidine (= N''-cyano-N-methyl-N'-[2-[[(5-methyl-1H-imidazol-4-yl)methyl] thio]ethyl]guanidine), which is a powerful antihistaminic drug (anti-H_2) (Fig. 2).   Highly soluble in water, weakly soluble in organic solvents, its extraction yield from body fluids is usually very low.   Through its nitrogen and sulfur atoms cimetidine gives

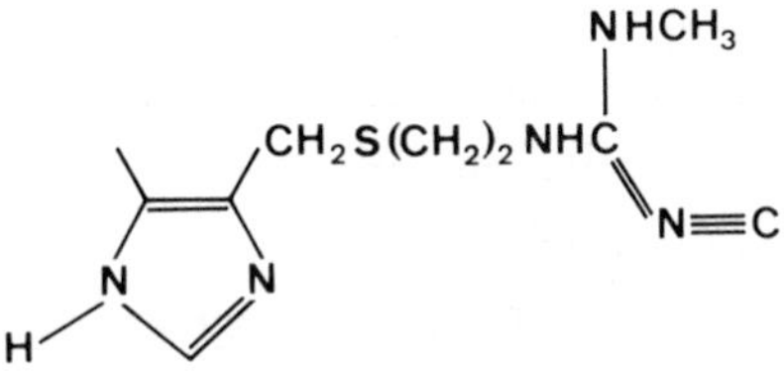

FIG. 2.   Cimetidine.

stable cationic chelates with various metal ions, e.g., Zn^{2+} and Cu^{2+}, which are soluble in water but insoluble in nonpolar solvents such as chloroform [7]. Using this chelate formation the extraction of cimetidine can be performed as follows.

In cacodylate buffer at pH $\sim$ 6 a copper salt is added to plasma or serum. The proteins precipitate and cimetidine remains in the aqueous phase; after mixing with chloroform and centrifuging, three phases can be separated:

1. A lower organic one, containing the lipids in chloroform solution
2. A medium semisolid one, consisting of precipitated proteins
3. An upper aqueous, clear one, containing the drug (and its metabolite sulfoxide) as a copper chelate

Cimetidine is extracted three times at pH 8.5 in ethylacetate from an aliquot of the upper phase (yield = 75%). After distillation of the solvent under nitrogen, the residue is taken up in the elution solvent and quantified by high-pressure liquid chromatography (UV detection at 228 nm) [8].

3.   INDIRECT CHELATION METHODS

3.1. Isoniazid

Isoniazid (= 4-pyridinecarboxylic acid hydrazide = isonicotinic acid hydrazide = INH) is a commonly used tuberculostatic drug. As early as 1959 Nielsch and Giefer proved that there is no relationship between the dose of INH and its concentration in blood because the patients are either "slow" or "rapid" acetylators. Acetylation, the first step of metabolism, inactivates the drug; the other metabolites are also inactive except the combinations of INH and ketoacids which are in equilibrium with their parent compounds [9].

The quantification of free INH can be performed by several methods in clinical pharmacology. The Schiff base formed between INH and salicylaldehyde has been used by Boxenbaum and Riegelman

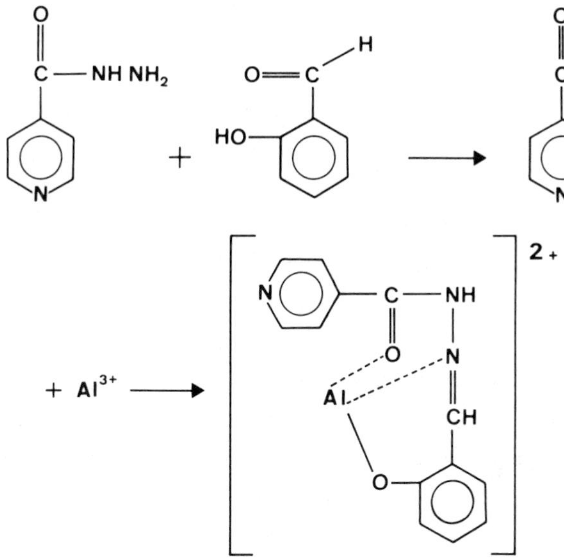

FIG. 3. Schiff base formation between isoniazid and salicylaldehyde followed by complex formation with Al^{3+}.

for such a purpose but the methodology is time consuming and its sensitivity is rather low [10]. We have noticed that this Schiff base gives highly fluorescent chelates with various metal ions, especially Al^{3+} (Fig. 3). After deproteinization by $Cd(OH)_2$, the Schiff base is obtained quantitatively at pH 1 in a few seconds; the aluminium chelate is formed around pH 5 and then extracted at this pH in amylalcohol. The fluorescence intensity is measured directly in this solvent. Technical details have been published elsewhere [11] and automatization is easily performed [12]. This highly sensitive and specific method is useful at the beginning of the therapy in order to determine for every patient the dose giving an active but nontoxic concentration of INH in blood (1-2 µg/ml).

3.2. Cannabinoids

3.2.1. *Introductory Remarks*

The characteristic feature of all these compounds is the aromatic
ring symmetrically substituted on carbon atoms 1, 3, and 5 (dibenzo-
pyran numbering). This structural unit exists in a simple compound,
olivetol, which is used for the synthesis of Δ9-tetrahydrocannabinol
(Δ9-THC). Symmetrical substitutions induce a characteristic reactivity, especially toward diazo reagents. As a classical rule a phenol
group induces electrophilic substitutions in the para position and,
to a smaller extent, in the ortho position(s). For example, in
olivetol, the two possible electrophilic substitutions involve a
carbon atom between a phenol group and the side chain, i.e., carbon
atoms 4 and 6 only (Fig. 4). In any case, the first electrophilic
substitution concerns the para function related to the first phenol
group, which is the ortho one related to the second phenol group.
Consequently, an o-hydroxydiazonium salt forms always an oo'-dihydroxy-
azo compound. To transform olivetol into such a compound, many diazo
reagents can be used but the simplest one derived from o-aminophenol
gives good results despite its low stability (photosensitivity).

The oo'-dihydroxyazo compounds give highly colored chelates
with Cu^{2+}, Ni^{2+}, Co^{2+}, Fe^{3+}, In^{3+}, Ga^{3+}, UO_2^{2+}, and Th^{4+}. Some of
them are only fluorescent (Al, In, Ga). The gallium chelates are
the most emissive; moreover, their emission wavelengths are in a
spectral range (595 nm) where few natural products are fluorescent.
These chelates are easily extracted from an aqueous medium by organic
solvents such as chloroform, methane dichloride, ethylacetate, and
aliphatic alcohols. Actually, we extract in n-amyl alcohol because
this alcohol provides the best compromise between aqueous solubility
and extractibility and the addition of dimethylformamide greatly
enhances the fluorescence of the gallium chelates [13].

Therefore, olivetol is considered as a model not only for
Δ9-THC and its metabolites, but also for most of the cannabinoids
(cannabinol, cannabidiol, cannabichromene, etc.). A sensitive and

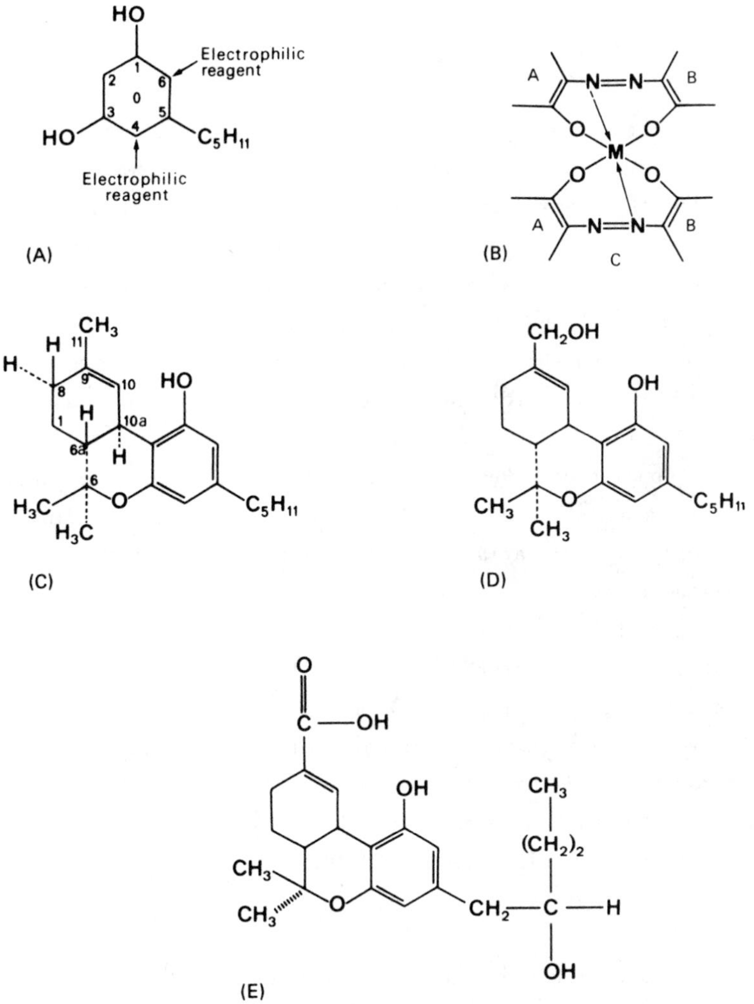

FIG. 4. Structure and reactivity of (A) olivetol, (B) proposed
structure of the metal chelate of oo'-dihydroxyazo compounds,
(C) Δ9-THC, (D) 11,16-dihydroxy-Δ-9-THC, (E) 11-carboxy-2'-hydroxy-
Δ9-THC.

specific method for identifying and quantifying cannabinoids has
been established from the above chemical considerations using the
fluorescence of the gallium chelates.

3.2.2. Optimization of Chelate Formation

1. Diazoreagent preparation. The diazonium salts of o-
aminophenol are sensitive to heat and light. With time the fluorescence of the blank solution increases rapidly due to autocoupling
of the reagent. To reduce this fluorescence, the diazonium salt
should be prepared in water-methanol medium at low temperature
($\sim 0°C$) in the dark and following standardized reaction conditions.
The diazo reaction should be stopped after 2 min with ammonium
sulfamate. The diazo reagent prepared in this way can be used for
at least 30 min.

2. Copulation. Resorcinol and its derivatives react easily
in 0.2 N aqueous solution of sodium hydroxide. In this medium,
Δ9-THC does not react at all with the mentioned diazonium salt:
its maximum reactivity is reached in 8-10 N NaOH. On the contrary,
cannabinol reacts better in dilute alkaline medium and it does not
react at all in concentrated alkaline solution (higher than 6 N).
After dissolving cannabinol and Δ9-THC in a few drops of methyl
alcohol and copulation in the proper medium, the reaction products
are extracted by amyl alcohol. After centrifugation, the organic
phase is washed with acetate buffer (pH 4.6), added to a gallium
salt solution containing also dimethylformamide to increase the
fluorescence intensity.

3.2.3. Quality Criteria of the Proposed Method

1. Sensitivity. The sensitivity is the smallest amount of
any compound that can be differentiated from zero. Taking into
account the mean emission value (± standard deviation of the blank),
this smallest amount has been chosen as giving a fluorescence intensity equal to the mean emission value + 3 SD (roughly 4 ng in a
sample of 2 ml of pure solution of Δ9-THC).

2.   Linearity has been checked on pure solutions of Δ9-THC
for the range of 10-1000 ng in the sample.

3.2.4. Applications

The proposed method can be used for identification of cannabinoids
in the natural products hashish, marijuana, and resin, in cigarettes
or butts after the extraction with hexan.   The cobalt chelate (pink
at pH 6) and the gallium chelate (red and fluorescent at pH 4.5) fit
best.

The cannabinoids excreted in urine are mainly in hydroxylated
or carboxylated forms [14] (Fig. 4).   For identification the extraction can be performed on Extrelut (Merck) using 4 ml of urine and
25 ml of hexan:ethylether (1:3 v/v).   For quantification, high-
pressure liquid chromatography and radioimmunoassay give better
results [15-17].   The cannabinoids present in blood are mostly
unmetabolized.   Therefore, after extraction and purification [17],
Δ9-THC may be quantified in blood till the third hour after intake
or smoking as long as the sample is large enough (plasma volume =
5 ml).

3.3. Colchicine

Only few publications have been devoted to colchicine metabolism;
this is probably due to its low concentration in body fluids.   We
have studied a method of quantification which can be used in emer-
gency because there are numerous and serious acute intoxications by
this alkaloid.

3.3.1.   Structure and Properties of Some
Colchicine Hydrazides

In 1957 Pesez [18] described a colored reaction based on the combination of colchicine with nicotinic (NHC) and isonicotinic hydrazides

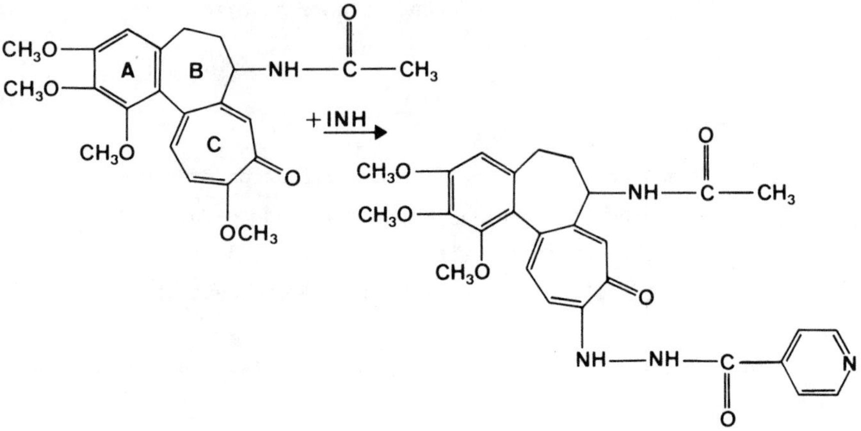

FIG. 5.   Reaction of colchicine and isonicotinic hydrazide (INH) to the isonicotinoyl hydrazide of colchicine.

(INHC), but the sensitivity ($\sim$10 μg/ml) was too low for the quantification in body fluids (Fig. 5). We have noticed that the reaction products, i.e., NHC and INHC, give chelates with various metal ions, especially Cu^{2+}, Al^{3+}, Ga^{3+}, and Sc^{3+}. The gallium chelates of the isonicotinic derivative are highly fluorescent and can be used for high-sensitivity quantification. In order to apply this chelation to measurements in body fluids and to optimize the working conditions, the following determinations were necessary.

3.3.2.   Acidity Constants of Colchicine Hydrazides

The competition existing in INHC or NHC solutions between H^+ and Ga^{3+} (or any other chelating cation) is revealed by the pH dependency of the visible spectra (Fig. 6) and by the pH dependency of the rate of chelate formation (Fig. 7). In addition, in homogeneous or heterogeneous medium, the fluorescence intensity increases from pH 1.8 to 3.5 and then decreases slowly, being zero at pH $\geq$ 6

INHC and NHC have three acidity constants which allow the explanation of the spectral modifications. The ligand L (NHC or INHC) exists in different forms depending on the pH:

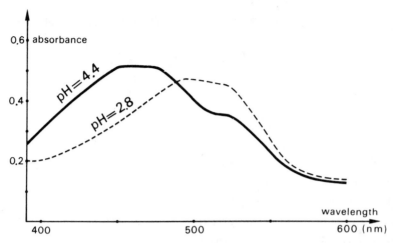

FIG. 6. The pH dependence of the spectra of the gallium chelate of INHC.

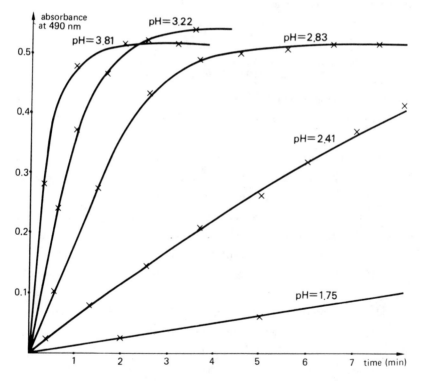

FIG. 7. The pH dependence of the rate of the chelate formation between Ga^{3+} and INHC. Reproduced from the Ph.D. thesis of M. Galliot [19].

$$LH \rightleftharpoons L^- + H^+ \qquad pK_{a/3} = 7.02 \text{ for INHC and } 7.15 \text{ for NHC}$$

$$LH_2^+ \rightleftharpoons LH + H^+ \qquad pK_{a/2} = 3.54 \text{ and } 3.29$$
$$\text{(protonation at the pyridine N)}$$

$$LH_3^{2+} \rightleftharpoons LH_2^+ + H^+ \qquad pK_{a/1} = 0.90$$
$$\text{(protonation on the tropolone)}$$

The competition observed in the pH range 0.5-2 (Fig. 6) is related to the tropolonic residue: with increasing pH the tropolonic cycle is more and more deprotonated allowing the substitution by Ga^{3+}. Also at pH below 3, the deprotonation of the pyridine nucleus is practically negligible as is the N fixation of any metal ion under these conditions. At pH about 3.2 there are two kinds of INHC derivatives, namely, the tropolonic ones, i.e., complexation occurs only at the tropolonic site, and those where the tropolonic site and also the pyridine N atom participate in complex formation.

3.3.3. Stoichiometry and Stability Constants of the Complexes [19]

The stability constants have been measured by using methods described elsewhere [3-5]. The stoichiometry is always of the type M/L = 1:1 and 1:2.

1. 1:1 Complexes. These are observed when the pH is above $pK_{a/2}$. The pyridine N atom and the tropolone hydrazide system are both involved in the formation of these complexes. Each ligand molecule is linked by two sites to two different metal ions: a first one by the tropolone hydrazide system and a second one by the N-heterocyclic atom. Such a structure leads to compounds of 1:1 stoichiometry with high molecular weights and a very low aqueous solubility.

2. 1:2 Complexes. When the pyridine N atom is protonated (pH $\lesssim pK_{a/2}$), coordination occurs only at the tropolone hydrazide site, the metal ion being hexacoordinated. In this case, the gallium complex carries a positive charge insuring the aqueous solubility. At pH 3.5-5.3 the pyridine N atom is deprotonated, but the tropolone hydrazide system is not yet negatively ionized. Each

ligand molecule is again involved in the chelate formation by its two sites but the metal ion is surrounded by four ligand molecules.

Infrared measurements support the proposed structures: in INHC, the 1620 cm^{-1} band corresponds to the C=O groups (tropolone and hydrazide), and in gallium-(INHC)$_2$, isolated from an acid medium at pH around 2.5, this band is not existing. In gallium-(INHC)$_2$, isolated at pH 4.5, this band is also missing and furthermore the 1060 cm^{-1} band, representative of NH deformation, is strongly reduced.

3.3.4. Application to Colchicine Quantification in Body Fluids

Taking into account the results mentioned above, colchicine can be quantified in:

1. Urine. The different steps are:

Extraction in the dark by CHCl$_3$

Condensation with isoniazid in alkaline medium (90°C, 12 min)

Extraction of the reaction products

Chelation with Ga^{3+}

Purification by chromatography on calibrated pores glass
     beads (CPG)

Fluorescimetry in amyl alcohol

2. Blood. The method can also be applied either to plasma or to red cells (where the concentration is roughly twice). The steps are the same as before. Details have been described elsewhere [20].

Up to now, the pharmacokinetic of colchicine was traced mainly by the use of ^{14}C-labeled samples [21-23]. The method we have described here has been applied during colchicine therapy and also during some acute intoxications.

3.3.5. Colchicine Therapy

The daily dose is usually 1 mg orally. After such an intake the concentration in blood can be followed for 2 hr; its maximum (roughly 10 ng/ml) is noticed around the 70th minute. In urine,

the quantification is easily performed for 12 hr and identification
for 24 hr. During a clinical trial, the urinary excretion has been
followed in 20 patients taking 1 mg in one dose. The excretion is
quick but always lower than 20% of the dose. Taking into account
the cumulated excretion and using the sigma minus method [24], the
mathematical analysis shows:

> An absorption phase ($t_{1/2}$ = 13 min)
>
> A first compartment ($t_{1/2}$ = 61 min), also previously
>   observed [25,26], and
>
> A second compartment, never observed before, whose half-life
>   ($t_{1/2} \sim$ 9 hr) is in good agreement with the biological
>   effects of colchicine [27]

3.3.6. Acute Intoxications

1. Oral intake up to 0.3 mg/kg body weight. In blood the
concentration is above 5 ng/ml for at least 12 hr. The urine excretion is precocious, i.e., 75% of the total excreted amount during the
first 8 hr. Nevertheless, the clearance (15-35 ml/min) is lower than
normal because collapsus is always associated. The urinary excretion
can be followed for more than 6 days.

2. Oral intake above 0.3 mg/kg body weight. The quantification
in blood is of great prognostic value. When the concentration is
above 30 ng/ml after the 10th hour, the case will probably be fatal.
If not, the concentration in blood can be followed up to the 48th
hour (c = 5 ng/ml). The graph "total quantity in the body versus
time" shows saturation for more than 6 days. The pharmacokinetic
parameters cannot be measured either in zero order or in first order
because delayed absorption and excretion are simultaneously observed.
Around the seventh day, the normal first order excretion rate reap-
pears. In any case the main excretion is performed by the feces:
70-90% of the intake dose during the first 20 hr.

ABBREVIATIONS AND DEFINITIONS

INHC	isonicotinoyl hydrazide of colchicine
NHC	nicotinoyl hydrazide of colchicine
NHC	nicotinoyl hydrazide of colchiceine
$\Delta 9$-THC	$\Delta 9$-tetrahydrocannabinol
$t_{1/2}$	half-life time

REFERENCES

1. A. S. Curry, *Br. Med. J.*, *1*, 354 (1964).

2. R. Bourdon, A. M. Nicaise, and J. Pollet, *Ann. Biol. Clin.*, *109*, 32 (1974).

3. P. Job, *Ann. Chim.*, *9*, 113 (1928).

4. S. Fronaeus, *Acta Chem. Scand.*, *5*, 139 (1951).

5. P. Souchay and J. Lefebvre, *Equilibres et réactivités en solution*, Masson, Paris, 1968, p. 27.

6. R. Bourdon and J. Hoffelt, *Ann. Biol. Clin.*, *38*, 351 (1980).

7. J. J. McCullough, W. K. McInnis, C. M. L. Lock, and R. Faggiani, *J. Am. Chem. Soc.*, *102*, 7782 (1980).

8. J. Hoffelt and R. Bourdon (in press).

9. W. Nielsch and L. Giefer, *Arzneim.-Forsch.*, *9*, 636 (1959).

10. H. G. Boxenbaum and S. Riegelman, *J. Pharm. Sci.*, *63*, 1191 (1974).

11. R. Bourdon, A. M. Nicaise, and J. Pollet, *Ann. Biol. Clin.*, *34*, 127 (1976).

12. M. Galliot, S. Perrault, and R. Bourdon (in press).

13. D. C. Freeman and C. E. White, *J. Am. Chem. Soc.*, *78*, 2678 (1956).

14. G. Nahas, W. Paton, and I. J. Heikkila, *Marihuana: Chemistry, Biochemistry and Cellular Effects*, Springer Verlag, New York, 1976, p. 169.

15. R. Bourdon, *J. Eur. Toxicol.*, *9*, 11 (1976).

16. P. L. Williams, A. C. Moffat, and L. J. King, *J. Chromatogr.*, *186*, 595 (1976).

17. J. M. Scherrmann, H. Hoellinger, N. H. Nam, and R. Bourdon, *Clin. Chim. Acta, 79,* 401 (1977).

18. M. Pesez, *Ann. Pharm. Fr., 15,* 630 (1957).

19. M. Galliot, Ph.D. thesis, Université René Descartes, Paris, 1979.

20. R. Bourdon and M. Galliot, *Ann. Biol. Clin., 34,* 393 (1976).

21. E. J. Walaszek, J. J. Kocsis, G. V. Leroy, and E. M. K. Geiling, *Arch. Int. Pharmacodyn. Ther., 125,* 371 (1960).

22. C. Boudene, F. Duprey, and C. Bohuon, *Biochem. J., 151,* 413 (1975).

23. D. Jarvie, J. Park, and J. Stewart, *Clin. Toxicol., 14,* 375 (1979).

24. J. G. Wagner, *Fundamentals of Clinical Pharmacokinetics,* Drug Intelligence Publications, Inc., Hamilton, Illinois, 1975.

25. S. L. Wallace and N. H. Ertel, *Metabolism, 22,* 749 (1975).

26. H. Halkin, S. Dany, M. Greenwald, Y. Shnaps, and M. Tirosh, *Clin. Pharmacol. Ther., 28,* 82 (1980).

27. R. Bourdon, M. Galliot, and J. M. Scherrmann, 2nd International Congress of Toxicology, Brussels, 1980.

Chapter 16

METAL COMPLEXES OF SULFANILAMIDES IN PHARMACEUTICAL ANALYSIS AND THERAPY

Auke Bult
Department of Pharmaceutical Analysis
and Analytical Chemistry
Subfaculty of Pharmacy
Gorlaeus Laboratories
State University
Leiden, The Netherlands

1. INTRODUCTION

Since the introduction of the first antibacterial active sulfanilamide derivative Prontosil in 1932 and the investigation of the
mechanism of action by the Nobel prize awarded Domagk [1,2], an
overwhelming number of sulfanilamides were synthesized and tested

for their biological activity. In the following years the well-
known sulfanilamides sulfanilamide (1936), sulfapyridine (1938),
sulfathiazole (1939), and sulfadiazine (1940) were introduced in
therapy. This class of compounds were the first effective chemotherapeutic agents to be employed systemically for the prevention
and cure of bacterial infections in humans. The advent of antibiotics has made large inroads on the popularity and fields of
usefulness of sulfanilamides. They continue, however, to occupy
an important although relatively small place in the therapeutic
armamentarium. Especially the management of ophthalmic infections
and infections of the urinary tract and gastrointestinal tract are
favored. The topical application of sulfanilamides is objectionable
because of the high incidence of sensitization and allergies.

The mechanism of action is based on the competitive antagonism
of PABA and the sulfanilamide (Woods-Fieldes theory). The sulfanilamide inhibits bacterial growth by preventing PABA from being incorporated into the folic acid molecule. Sensitive microorganisms (a
wide range of gram-positive and gram-negative organisms) are those
that must synthesize their own folic acid, the other bacteria are
not affected. Only in high concentration is a bactericidal effect
reached (urinary and gastrointestinal tract).

The general structural formula of this class of compounds and
the most important representatives are given in Fig. 1. Mostly R_1
is a heteroaromatic substituent. Incidentally, a hydrophilic substituent R_2 is introduced at the aromatic $^4NH_2$ group (succinyl,
phthalyl). Now absorption from the gastrointestinal tract is prevented. Sulfanilamides exhibit an amido-imido tautomerism [3]. The
equilibrium position depends on the character of R_1, the physical
state (solid, solution), and the solvent polarity.

In general sulfanilamides are badly soluble in water and apolar
solvents e.g., diethylether and chloroform, and freely soluble in
more polar solvents, e.g., acetone and dimethyl sulfoxide. The solubility is much better in acidic and alkaline solution because of pro-
tonation of $^4NH_2$ and deprotonation of the acidic ^1NH group, respectively. The acid pK_a values of ^1NH range from about 6 to 8 [4].

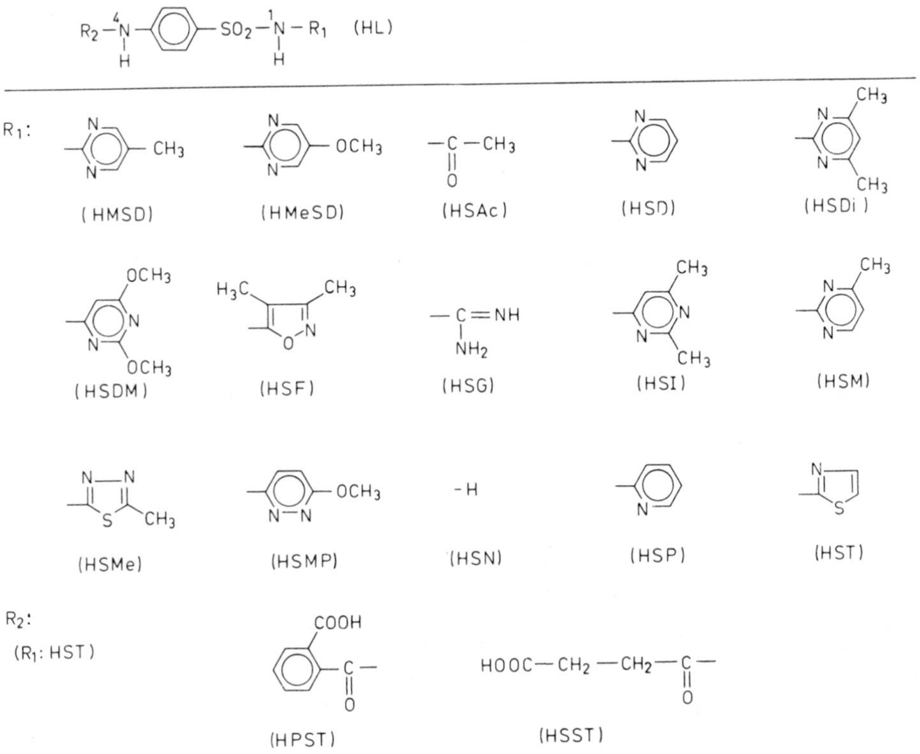

FIG. 1. General structure and microbiological active representatives of the sulfanilamides.

The analytical chemistry of the sulfanilamides has got much attention. Many surveys are available [5,6].

Recently, the interest in sulfanilamides is renewed by reason of the application of their metal compounds in topical burn therapy. The silver compounds receive special attention in Sec. 5.

2. SYNTHESIS AND STRUCTURE OF METAL SULFANILAMIDES

2.1. Synthesis

In principle it is possible to prepare metal sulfanilamide compounds with most sulfanilamides and most metal ions. However, the compounds with transition elements and main elements of groups Ib and IIb are most frequently studied. There are described four different types of compounds. Their routes of preparation are:

(A)

$$MX_n + mHL \xrightarrow{\text{solvent}} M(HL)_m X_n \downarrow \qquad (1)$$

The sulfanilamide HL is present as neutral ligand in the compound formed. X is present as counterion (NO_3^-) or coordinated to M (Cl^-, I^-, SCN^-, acetate). The solvent used is methanol, acetone, ether, or water (pH: 4-5). Examples are M = Ag(I) [7], Bi(III) [8], Hg(II) [9,10], Cu(II) [11], Cd(II), Co(II), Mn(II), Ni(II), Zn(II) [12,13].

(B)

$$MX_m + mHL \xrightarrow[\text{pH 7-10}]{\text{water}} ML_m \downarrow + mHX \qquad (2)$$

or

$$MX_m + mNaL \xrightarrow{\text{water}} ML_m \downarrow + mNaX \qquad (3)$$

HL has been deprotonated. For the deprotonation to L^- a slightly alkaline medium (pH: 7-10) is needed [Eq. (2)]. Alternatively the corresponding sodium sulfanilamide can be used for the preparation [Eq. (3)]. Examples are M = Ag(I) [14,15], Cu(II) [16], Cd(II), Co(II), Mn(II), Fe(III), Ni(II), Zn(II) [17,18]. An alternative preparation of the compounds ML_m [M = As(III), Bi(III), Fe(III), Sb(III), Sn(II), Zn(II)] was described by Ruskin and Pfaltz [19].

$$Ca \text{ (metal)} + 2CH_3OH \longrightarrow Ca(OCH_3)_2 + H_2 \uparrow \qquad (4)$$

$$Ca(OCH_3)_2 + 2HL \xrightarrow{\text{methanol}} CaL_2 + 2CH_3OH \qquad (5)$$

$$mCaL_2 + 2MCl_m \xrightarrow{\text{methanol}} 2ML_m \downarrow + mCaCl_2 \qquad (6)$$

The corresponding HgL_2 compounds prepared according to this procedure are also described [20].

(C)

$$ML_2 + 2HCl \xrightarrow[\text{pH 4-5}]{\text{water}} M(HL)_2 Cl_2 \xrightarrow[\text{pH 1-1.5}]{\text{water}} (H_2L)_2[MCl_4] \qquad (7)$$

HL is protonated to H_2L^+ and present as counterion. Examples are MCl_4 = $CuCl_4$, $CoCl_4$, $CdCl_4$, $NiCl_4$ [12,17,18].

(D)

$$ML_m + nY \longrightarrow ML_m Y_n \qquad (8)$$

Y is a suitable solvent or a solution of Y, e.g., NH_3, morpholine, imidazole, pyridine, n = 2-4. Examples are M = Ag(I) [21], Cd(II), Co(II), Cu(II), Mn(II), Ni(II), Zn(II) [12,17].

An exhaustive review on the synthesis of metal sulfanilamide compounds recently appeared [22].

2.2. Structure

It is hardly possible to give a generally valid treatise on the structures of metal sulfanilamide compounds. The nature of the metal ions (a class and b class metals with different preference to the various donor atoms in HL) strongly varies and also the molecular structures of the various sulfanilamides are rather different. From Fig. 1 is clear that the donor sites mostly present in the substituent R_1 will have a strong influence on the character of the metal compounds formed. It is necessary to make a separate study of every individual compound.

The number of well-documented studies on the structures of metal sulfanilamide compounds is very limited. This knowledge is important to get insight into the selectivity of identification tests and to evaluate quantitative determinations based on metal complex formation. In this section a short review of the structural features is presented.

Compounds with composition $M(HL)_mX_n$. Gulko et al. [23] and Alléaume et al. [24] made an x-ray analysis of $Pd(II)(HSG)_2X_2$ (X = Cl; Br). Pd(II) is four-coordinate (square planar) and interacts with HSG through the aromatic $^4NH_2$. Ferrari et al. [8] studied $Bi(III)(HSMP)_3Cl_3$ by x-ray analysis. Bi(III) was nine-coordinated and each of the three HSMP acts as a bidentate. A schematic representation of the structure is given in Fig. 2, structure I. Most studies of $M(HL)_m$ are based on IR analysis and in case of transition elements eventually supplemented by ligand field spectra, μ_{eff} measurements, and ESR. Coordination of M through $^4NH_2$ and $-SO_2-^1NH-$ results in a negative shift of the stretching frequencies of $^4NH_2$,

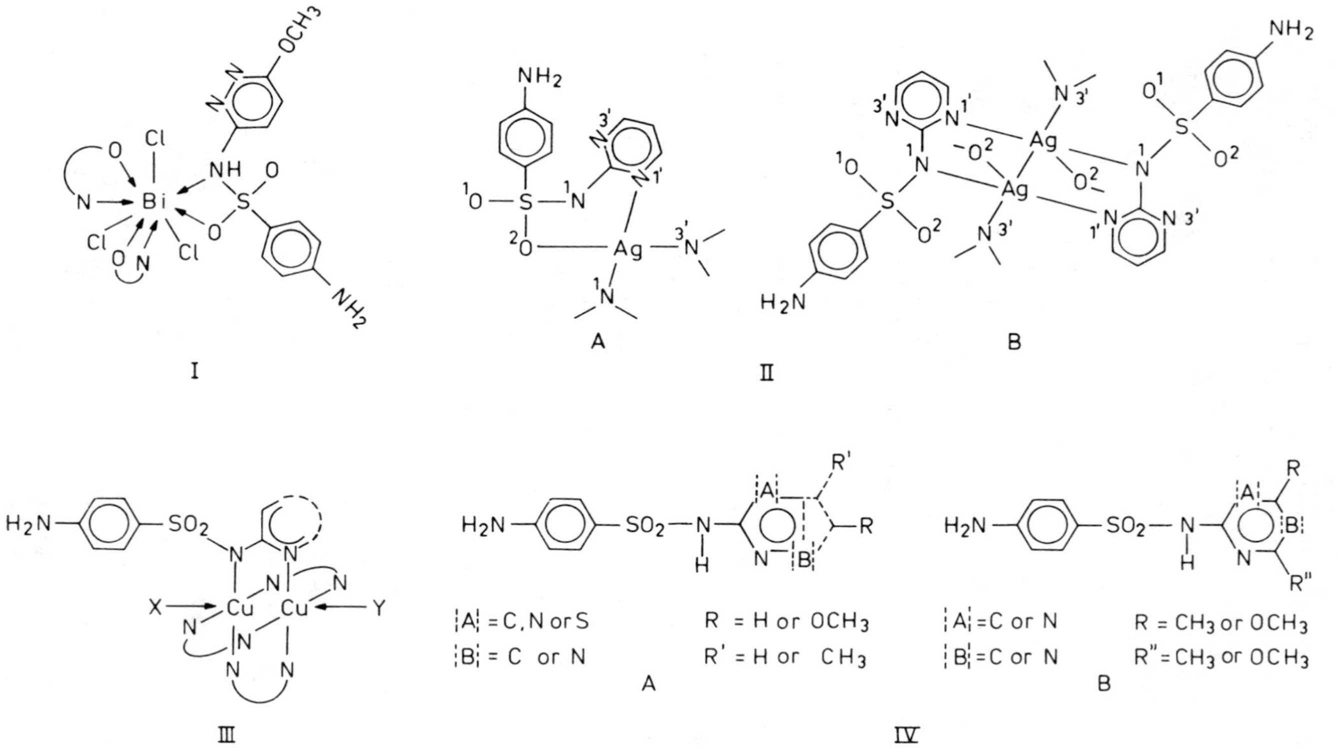

FIG. 2. Structures of some metal sulfanilamide compounds: I, based on a crystal structure of Bi(HSMP)$_3$Cl$_3$ [8]; II, structural representation of AgSD according to (A) Cook and Turner [27], (B) Baenziger and Struss [26]; III, structural representation of dimeric Cu$_2$L$_4$; IV, common structural element of the sulfanilamides forming (A) dimeric Cu$_2$L$_4$ and (B) polymeric (CuL$_2$)$_n$ compounds.

[1]NH, and SO_2 [22]. Mostly it is possible to locate in this way one or two donor sites in HL, but the location of donor sites in the heteroaromatic substituent R_1 of HL is difficult. Well-documented studies are made of $Cu(II)(HL)_2(acetate)_2$, HL = HSD, HSDi, HSM, HST [11] and $Cu(II)(HL)_2Cl_2$; HL = HSD, HSG, HSM, HSN, HST [25]. In both types of compounds Cu(II) is four-coordinated (planar) and interacts through [4]NH_2. On the contrary, Shukla and Bhatia [9] described $Hg(HL)_1Cl_2$ (HL = p-toluene, N-alkylsulfonamides) with a structure similar to Fig. 2, structure I.

   Compounds with composition ML_m *and* ML_mY_n. Prior to the formation of ML_m, HL is deprotonated: $R'-SO_2-$[1]$NH-R_1 \xrightarrow{-H^+} R'-SO_2-N-R_1^-$. L has a negative charge and his donor properties to M are better as compared with HL. As a result coordination with M mostly occurs through the $-SO_2-N-R_1$ group. In case R_1 is a heteroaromatic substituent a partial transfer of electron density to R_1 by resonance takes place and now R_1 is often involved in coordination. The best documented study on this type of compounds is AgSD (silver sulfadiazine). The results of two independent x-ray analyses are given in Fig. 2, structures IIA and IIB [26,27]. In structure IIA Ag(I) is four coordinated (tetrahedral) and surrounded by three different SD molecules and in structure IIB Ag(I) is five-coordinated (trigonal bipyramidal) and surrounded by four different SD molecules. Actually in both structures the Ag-Ag distance is equal (2.92 Å). The donor atoms are the same in both structures; AgSD is a polymer. Bult and Klasen [14,15] studied an extensive series of 18 AgL compounds by IR. They concluded to two types of AgL: amido-Ag and imido-Ag. In amido-Ag one of the donor sides is O of SO_2^- (like in AgSD). The imido-Ag structure is common. An Ag ← [4]NH_2 interaction is found scarcely. Compounds of the type $[AgL_2]Ag$ are also described [28].

   The compounds CuL_2 have also been studied [16]. Two types of compounds were found: dimeric and polymeric. The dimeric compounds had the composition Cu_2L_4 (L = MeSD, SD, SM, SMP, SP, ST) and Cu(II) was four- or five-coordinated (planar- or square-based pyramidal, respectively).

In structure III of Fig. 2 is given a representation of the
structures. The colors of these compounds were purple to red-brown.
The polymeric compounds $(CuL_2)_n$ (L = SDi, SDM, SI) were yellow or
ochre. Their structural evidence is incomplete. The common structural element of the sulfanilamides forming dimeric compounds (Fig.
2, IVA) and polymeric compounds (Fig. 2, IVB), respectively, is
shown. The formation of dimeric compounds is preferred but the
presence of group R'' in α position relative to N (Fig. 2, IVB) prevents this formation due to steric hindrance; as a result polymeric
compounds are formed. Sulfanilamides without a heteroaromatic R_1
substituent do not form CuL_2 compounds, e.g., L = SAc, SN.

A compound of type ML_mY_n is described by Bult [29]. In
$Co(ST)_2(R-NH_2)_2$ (R-NH$_2$ = isobutylamine), Co(II) is four-coordinated
(tetrahedral) and binds ST through the N atom of thiazole.

3. IDENTIFICATION TESTS BASED ON METAL COMPLEX FORMATION

The methods of preparation discussed in Sec. 2.1 are in principle
suitable for application in qualitative and quantitative analysis of
sulfanilamides. The ultimate utility in qualitative analysis depends
on the possibility to detect the reaction products and the selectivity
of the reaction. The detection can be based on the observation of a
color, a (colored) precipitate, or a crystalline product. The selectivity of a reaction is based on the nature of the color(s), change
of color with time, or the crystal form of the reaction product. A
color reaction is useful if it is very selective for a group of analog
compounds (like the sulfanilamides) or can be used for differentiation
within the group of analogs. Crystal tests mostly have a high degree
of selectivity.

A frequently used identification test for sulfanilamides is
their color formation with Cu(II) salts. This test is included in
the modern pharmacopoeias (Ph.Eur.I, BP 1980, USP XX) and handbooks
on pharmaceutical analysis [5,6].

The most frequently applied modification of the test uses an aqueous alkaline solution of HL and a $Cu(II)SO_4$ solution as reagent. There are various (slightly) different procedures in use. Sometimes the reaction conditions are unfavorable [30]. The best results are obtained if the pH of the reaction solution is between 8 and 11, the concentration of Cu(II) in the reaction solution is about 0.1 mmol/ 10 ml, and the ratio of [HL]/[Cu] is about 1 [30].

A satisfactory procedure is to dissolve about 50 mg of the sulfanilamide in a mixture of 8 ml of water and 2 ml of 0.1 N sodium hydroxide and add 0.2 ml of copper sulfate solution (12.5% w/v $CuSO_4 \cdot 5H_2O$). The color formation is observed during 5 min.

The selectivity of the test within the group of sulfanilamides is based on differences in color and change of color with time. The main colors are purple to red-brown, green, ochre, and yellow. In practice a wide variety of intermediate colors is found. The best procedure is the comparison of the developed color with the colors of the blank and the reference compound. This identification test is based on the formation of a product with composition CuL_2. As already mentioned in Sec. 2.2, two types of CuL_2 exist: dimeric and polymeric. A prerequisite for the formation of CuL_2 was the presence of a heteroaromatic R_1 substituent in HL. The variation in R_1 combined with the two types of geometry of CuL_2 forms the basis of the selectivity of the test. A special application of this test is its use as a microcrystal test [31] and for the visualization of HL on TLC plates and paper chromatograms [32]. A modification of the described test is the reaction of HL with an aqueous Cu(II)-amine solution (amine = Y = pyridine, butylamine). The reaction products with composition CuL_2Y_{2-3} are green or brown and extractable with chloroform [33] (Pharmacopoeia Japonica 1973).

Very similar to the Cu(II)-amine test is the reaction of HL with cobaltous acetate in methanol. The color is developed on addition of an aliphatic amine like cyclohexylamine. The products formed have the composition CoL_2Y_2 and are tetrahedral (violet) or octahedral (pink-red) depending on the nature of the R_1 substituent. In Sec. 2.2

was given an example of such a compound. The selectivity of this
test is subordinate to the Cu(II) test [34].

A very selective procedure for the identification of HL is the
microcrystal test. There are many reagents in use, e.g., $AuBr_4^-$,
$AuCl_4^-$, HgI_4^{2-}, PbI_4^{2-}, and $PtBr_6^{2-}$ [5]. The sensitivity is about 1:100
or better. The nature of the crystalline products is not clear.

4.   QUANTITATIVE ANALYSIS BASED ON METAL COMPLEX FORMATION

Complex formation is one of the basic methods for the quantitative
analysis of HL [6]. In principle the reactions discussed in Sec. 2.1
are useful. However, the practical utility is governed by the following conditions: (1) the reaction should proceed reasonably far toward
completion, (2) the reaction must be such that it can be described by
a balanced chemical equation; this implies the absence of side reactions, (3) a satisfactory end point detection is required, (4) the
whole procedure must be reproducible. The importance of condition 1
is demonstrated by the following: HSN cannot be determined by complexometric titration with Ag(I) in contrast to most other HL compounds. The conditional stability constant of AgSN is too low
(log K = 0.96) compared with the normal values (log K > 2.8) [35].

Most of these methods are based on the formation of ML_m or
ML_mY_n. They are not generally applicable to the whole group of HL
compounds. Compounds HL with a heteroaromatic R_1 substituent give
the best results. The quantitative methods are based on the general
reaction:

$$MX_m + mHL \ (+ \ nY) \longrightarrow ML_m \ (ML_mY_n) + mHX \qquad (9)$$

The quantification is based on one of the following principles:

1.   Gravimetric analysis of ML_m:AgL [37] or dissolution of
     ML_m followed by complexometric titration of M: CuL_2Py_{2-3},
     titrant: EDTA; indicator: xylenolorange [43].

2.   Direct complexometric titration with a suitable indication
     method: AgL, ind: dichromate [37]; AgL, ind: thermometri-

cally [40]; AgL, ind: potentiometrically using a modified
graphite or silver ion-selective electrode [41]; CuL_2,
ind: potentiometrically using a modified graphite or copper
ion-selective electrode [41].

3. Determination of excess MX_m by complexometric back titration: Ag(I), titrant: SCN, ind: Fe^{3+} [36]; Ag(I), titr:
   EDTA, ind: $Ni(CN)_4$/murexide [38]; Cu(II), titr: EDTA, ind:
   PAN [42]; Hg(II), titr: EDTA, ind: potentiometrically using
   a liquid membrane Hg(II)-sensitive electrode [44].

4. Determination of excess MX_m by atomic absorption spectrometry (AAS): Ag(I) and Cu(II) [41].

5. Acidimetric titration of the liberated HX: ind: Phenol Red
   [39].

In Table 1 are summarized the relevant data of the mentioned methods.
Most methods are suitable for the analysis of pharmaceutical formula-
tions containing HL.

5.   PHARMACOCHEMICAL AND THERAPEUTIC ASPECTS OF METAL SULFANILAMIDE COMPOUNDS

At present some metal sulfanilamides get much attention owing to
their antimicrobial activity. Especially the Ag(I), Zn(II), and
Ce(III) compounds of sulfadiazine (SD) seemed to be suitable for the
treatment of burns. AgSD is now available as a 1% cream (Silvadene,
Flammazine).

The antimicrobial activity of various metal ions is well known,
e.g., Al, Ag, Bi, Cd, Ce, Cu, Hg, Zn [45]. For instance, the Ag(I)
ion is more or less active against both gram-positive and gram-
negative bacteria, fungi, and viruses. Therefore a 0.5% $AgNO_3$ solution was in use in burn treatment since decades. Disadvantages of
this therapy were among others the frequent application of the solution, the disturbance of the electrolyte balance of the patient
(precipitation of AgCl), and the inconvenient black staining as a
result of reduction to metallic silver.

TABLE 1

Analytical Methods for Sulfanilamides Based on Metal Complex Formation

Complexing ion	Method*	Av. recovery ± SD or error (%)	Determined sulfanilamides	Ref.
Ag	c	99.6 ± 0.2	HSD, HSDi, HSM, HSP, HSST, HST	[36]
Ag	a	99.5-100.5	HSD, HSDi, HSM, HSP, HSST, HST, HPST	[37]
Ag	b	99.5-100.0	HSD, HSDi, HSM, HSP, HSST, HST, HPST	[37]
Ag	c	99.0-100.4 ± 0.3-1.2	HPST, HSN, HSP, HSST, HST	[38]
Ag	e	101.2 ± 1.6	HST	[39]
Ag	b	98.0-102.0 ± 1.0	HSD, HSDi, HSM, HSMe, HSP, HST	[40]
Ag	b	99.7 ± 0.2	HMeSD, HSD, HSDi, HSM, HSMP, HSP, HST	[41]
Ag	d	99.4 ± 0.3	HMeSD, HSD, HSDi, HSM, HSMP, HSP, HST	[41]
Cu	c	99.1-99.8 ± 0.6	HSD, HSDi, HSM, HSP, HST	[42]
Cu	a	99.1-99.8 ± 0.4	HPST, HST	[43]
Cu	b	99.7 ± 0.2	HMeSD, HSD, HSDi, HSM, HSMP, HSP, HST	[41]
Cu	d	99.4 ± 0.3	HMeSD, HSD, HSDi, HSM, HSMP, HSP, HST	[41]
Hg	c	99.2 ± 0.4	HMeSD, HSF, HST	[42]

*Method: (a) gravimetric analysis, (b) direct complexometric titration, (c) complexometric back titration of excess reagent, (d) AAS spectrometry of excess reagent, (e) acidimetric titration of liberated acid.

The introduction of AgSD (1968) in this therapy represented important progress. The clinical results are good and the disadvantages of $AgNO_3$ are absent [46]. As already mentioned in Sec. 1, the topical application of sulfanilamides is objectionable because of the high incidence of sensitization. The very low water solubility of AgSD (2 μg/ml) results in a negligible sensitization: <0.1% [46]. Later on also $Zn(SD)_2$ [47] and $Ce(SD)_3$ [48] were recommended as antimicrobial agents in burn therapy. However, these two compounds are in fact physical mixtures of HSD and the corresponding metal hydroxides [49].

The mechanism of action of AgSD in relation to $AgNO_3$ and HSD has gotten much attention [50-53]. The antimicrobial active part of AgSD is Ag. The activity of Ag is based on the alteration of the mesosomal function of the microbial cell [52]. The inhibition of the bacterial growth is related quantitatively to the binding of Ag to microbial DNA [54]. This conclusion is confirmed by the finding that AgSD is not antagonized by PABA (Sec. 1) [51]. AgSD and $AgNO_3$ show a different membrane interaction with the microbial cell [50]. In the case of AgSD the whole molecule seems to be involved in the interaction; in the case of $AgNO_3$ the silver ion is active. The exact role of SD in the mechanism of action of AgSD is unclear but important as was demonstrated by the absence of the side effects seen with $AgNO_3$.

Some attempts have been made to get insight into the role of SD in AgSD in relation to other silver sulfanilamides [35,51,55-58]. From a series of nine AgL compounds were measured the acid dissociation constants (pK_a) of the sulfanilamides, the conditional stability constants (log K), and the solubility of the AgL compounds [35,57]. In Table 2 are summarized these physical data together with their biological data. There does not exist a linear relationship between one of the physical data sets and the in vivo activity (as percent mortality of mice with 30% burn and infected with *Pseudomonas aeruginosa*). The correlation coefficients r are not significant (Table 2). The in vitro activity of the nine AgL compounds is about the

TABLE 2

Biological and Physical Data of the
Silver Sulfanilamides (AgL)

AgL	% Mortality[a]	pK_a (HL)[b]	Log K (AgL)[b]	Solubility AgL (μg/ml)[c]
AgSD	25.6	6.46	3.57	0.50
AgST	40	7.19	3.97	0.10
AgSI	60	7.47	3.35	1.10
AgSMe	65	5.63	3.54	1.00
AgSN	77	10.57	0.96	7.00
AgSM	90	6.88	3.80	0.60
AgSP	75	8.57	2.82	1.10
AgMSD	80	6.68	3.58	1.90
AgMeSD	80	6.67	3.41	0.60
r[d]		0.07	0.21	0.19

[a]Refs. 55, 57, 59.

[b]Log K is the conditional stability constant at pH = 7.4 and ionic
strength 0.1 mol/liter (25°C) [35].

[c]Solubility in water calculated as the concentration of silver ions
in μg/ml.

[d]r: correlation coefficient between % mortality and the physical
data set.

same. Depending on the type of microorganism, the MIC values range
from 6.3 to 200 μg/ml [58]. This is far below the water solubility
of AgL. From this point of view it is hard to understand how a
therapeutic concentration of AgL is reached in the wound environs.
Another difficult point is the distinct behavior within the series
of AgL compounds under in vivo and in vitro experimental conditions.
The moderate value of the conditional stability constant of AgSD
(3.57) is responsible for the photostability of the compound and the
different mechanism of action from $AgNO_3$.

6. GENERAL CONCLUSIONS

In the literature the synthesis of metal sulfanilamide compounds has
received much attention but the structure elucidation of these compounds is backward. The applications of metal complex formation in
pharmaceutical analysis are mostly of older date. However, combination of these basic methods with modern instrumental methods of end
point indication creates new opportunities for quantitative analytical
application [41,44]. The therapeutical application of metal sulfanilamides is new. Consequently, the knowledge about the mechanism of
action and medicinal chemistry aspects is incomplete and there is
much work left to do in this field. For example, recently two more
water-soluble antimicrobial-active derivatives of AgSD were described
[21].

ABBREVIATIONS

EDTA	ethylenediaminetetraacetic acid
HL	sulfanilamide derivative
HMSD	5-methylsulfadiazine
HMeSD	5-methoxysulfadiazine
HPST	phthalylsulfathiazole
HSAc	sulfacetamide
HSD	sulfadiazine
HSDi	sulfadimidine
HSDM	sulfadimethoxine
HSF	sulfafurazole
HSG	sulfaguanidine
HSI	sulfisomidine
HSM	sulfamerazine
HSMe	sulfamethizole
HSMP	sulfamethoxypyridazine
HSN	sulfanilamide

HSP	sulfapyridine
HSST	succinylsulfathiazole
HST	sulfathiazole
ind.	indicator
M	metal cation (various valencies)
MIC	minimum inhibitory concentration
PABA	p-aminobenzoic acid
L, MSD, MeSD, etc.	anion of HL, HMSD, HMeSD, etc.

REFERENCES

1. G. Domagk, *Dt. Med. W. Schr.*, *61*, 250 (1935).

2. G. Domagk, *Dt. Med. W. Schr.*, *61*, 829 (1935).

3. A. Bult and H. B. Klasen, *Pharm. Weekblad*, *113*, 665 (1978).

4. H. Sakurai and T. Ishimitsu, *Talanta*, *27*, 293 (1980).

5. E. G. C. Clarke (ed.), *Isolation and Identification of Drugs*, Pharmaceutical Press, London, Vol. 1 (1978), Vol. 2 (1978).

6. S. Ebel, *Handbuch der Arzneimittel-Analytik*, Verlag Chemie, Weinheim, 1977, p. 191ff.

7. K. K. Narang and J. K. Gupta, *Curr. Sci.*, *45*, 744 (1976).

8. M. B. Ferrari, L. Calzolari Capacchi, L. Cavalca, and G. F. Gasparri, *Acta Crystallogr.*, *Sect. B*, *28*, 1169 (1972).

9. J. S. Shukla and P. Bhatia, *J. Inorg. Nucl. Chem.*, *36*, 1422 (1974).

10. G. S. Misra and M. V. Ramakrishna Rao, *J. Indian Chem. Soc.*, *54*, 1013 (1977).

11. K. K. Narang and J. K. Gupta, *Transition Met. Chem.*, *2*, 83 (1977).

12. P. V. Gogorishvili, M. G. Tskitishvili, R. I. Machkhoshvili, and Y. Y. Kharitonov, *Zh. Neorg. Khim.*, *20*, 1420 (1975), via *Chem. Abstr.*, *83*, 90119 m (1975).

13. M. G. Tskitishvili, P. V. Gogorishvili, R. I. Machkhoshvili, and N. B. Zhorzholiani, *Izv. Akad. Nauk. Gruz SSR, Ser. Khim*, *5*, 295 (1979), via *Chem. Abstr.*, *93*, 18.270 ω (1980).

14. A. Bult and H. B. Klasen, *J. Pharm. Sci.*, *67*, 284 (1978).

15. A. Bult and H. B. Klasen, *Arch. Pharm. (Weinheim)*, *311*, 855 (1978).

16. A. Bult, J. D. Uitterdijk, and H. B. Klasen, *Transition Met. Chem.*, *4*, 285 (1979).

17. P. V. Gogorishvili and M. G. Tskitishvili, *Issled. Obl. Khim. Kompleksn. Prostykh. Soedin Nek: Perekhodnykh. Redk. Met.*, *3*, 5 (1978), via *Chem. Abstr.*, *90*, 214.469 s (1978).

18. M. G. Tskitishvili, P. V. Gogorishvili, M. V. Chrelashvili, and A. E. Shvelashvili, *Izv. Akad. Nauk. Gruz SSR, Ser. Khim*, *5*, 13 (1979), via *Chem. Abstr.*, *92*, 163.917 t (1980).

19. S. L. Ruskin and M. Pfaltz, *Am. J. Pharm.*, *163* (1947).

20. R. Dolique and P. Mas, *Trav. Soc. Pharm. Montpellier*, *13*, 87 (1953).

21. A. Bult and H. B. Klasen, *Arch. Pharm. (Weinheim)*, *313*, 1016 (1980).

22. A. Bult, *Pharm. Weekblad, Sci. Ed.*, *3*, 1 (1981).

23. A. Gulko, W. F. Rittner, G. Ron, A. Weissman, and G. Schmuckler, *J. Inorg. Nucl. Chem.*, *33*, 761 (1971).

24. M. Alléaume, A. Gulko, F. H. Herbstein, M. Kapon, and R. E. Marsch, *Acta Crystallogr., Sect. B*, *32*, 669 (1976).

25. K. K. Narang and J. K. Gupta, *Indian J. Chem.*, *13*, 705 (1975).

26. N. C. Baenziger and A. W. Struss, *Inorg. Chem.*, *15*, 1807 (1976).

27. D. A. Cook and M. F. Turner, *J. Chem. Soc., Perkin Trans.*, *2*, 1021 (1975).

28. A. Bult and H. B. Klasen, *Arch. Pharm. (Weinheim)*, *313*, 72 (1980).

29. A. Bult, *Pharm. Weekblad*, *111*, 385 (1976).

30. A. Bult, J. D. Uitterdijk, and H. B. Klasen, *Pharm. Weekblad, Sci. Ed.*, *1*, 101 (1979).

31. A. Goudswaard, *Pharm. Weekblad*, *95*, 487 (1960).

32. M. I. Walash and S. P. Agarwal, *J. Pharm. Sci.*, *61*, 277 (1972).

33. Ch. Lapière, *J. Pharm. Belg.*, *3*, 17 (1948).

34. A. Bult and H. B. Klasen, *Pharm. Weekblad*, *109*, 389 (1974).

35. G. J. Boelema, A. Bult, H. J. Metting, B. L. Bajema, and D. A. Doornbos, *Pharm. Weekblad, Sci. Ed.*, *4*, 38 (1982).

36. P. L. de Reeder, *Anal. Chim. Acta*, *10*, 413 (1954).

37. L. Kum-Tatt, *Analyst*, *82*, 185 (1957).

38. C. Hennart, *Chim. Anal.*, *44*, 8 (1962).

39. W. Friedrich, C. Gehringer, and K. Mubarak, *Dt. Apoth-Ztg*, *110*, 1503 (1970).

40. L. S. Barke and J. K. Grime, *Analyst, 98,* 452 (1973).

41. S. S. M. Hassan, and M. H. Eldesouki, *J. Assoc. Off. Anal. Chem., 64,* 1158 (1981).

42. H. Abdine and W. S. Abdel Sayed, *J. Pharm. Pharmacol., 14,* 761 (1962).

43. W. Poethke and C. Jaekel, *Arch. Pharm. (Weinheim), 296,* 627 (1963).

44. M. Ionescu, S. Cilianu, A. A. Bunaciu, and V. V. Cosofret, *Talanta, 28,* 383 (1981).

45. T. D. Luckey and B. Venugopal, *Metal Toxicity in Mammals,* Vol. 1, Plenum Press, New York, 1977, p. 10ff.

46. C. L. Fox, in *Recent Advances in Dermatopharmacology,* Spectrum, New York, 1977, p. 41ff.

47. C. L. Fox, Sh. M. Modak, and J. W. Stanford, *Surg. Gynecol. Obstet., 42,* 553 (1976).

48. C. L. Fox, Sh. M. Modak, and J. W. Stanford, *Burns, 4,* 233 (1978).

49. A. Bult, N. Hulsing, and J. W. Weyland, *J. Pharm. Pharmacol., 33,* 171 (1981).

50. J. E. Coward, H. S. Carr, and H. S. Rosenkranz, *Antimicrob. Ag. Chemother., 3,* 621 (1973).

51. C. L. Fox and Sh. M. Modak, *Antimicrob. Ag. Chemother., 5,* 582 (1974).

52. H. S. Rosenkranz and H. S. Carr, *Antimicrob. Ag. Chemother., 2,* 367 (1972).

53. M. S. Wysor and R. E. Zollinhofer, *Pathol. Microbiol., 39,* 434 (1973).

54. Sh. M. Modak and C. L. Fox, *Biochem. Pharmacol., 22,* 2391 (1973).

55. R. U. Nesbitt and B. J. Sandmann, *J. Pharm. Sci., 67,* 1012 (1978).

56. A. Bult, B. L. Bajema, H. B. Klasen, H. J. Metting, and C. L. Fox, *Pharm. Weekblad, Sci. Ed., 3,* 79 (1981).

57. A. Bult, unpublished results.

58. J. J. M. v. Saene, A. Bult, and C. F. Lerk, *Pharm. Weekblad, Sci. Ed., 5,* 61 (1983).

59. J. W. Stanford, B. W. Rappole, and C. L. Fox, *J. Trauma, 9,* 377 (1969).

Chapter 17

BASIS FOR THE CLINICAL USE OF GALLIUM AND INDIUM RADIONUCLIDES[*]

Raymond L. Hayes and Karl F. Hübner
Medical and Health Sciences Division
Oak Ridge Associated Universities
Oak Ridge, Tennessee

---

[*]This article is based on work supported by contract number DE-ACO5-760R00033 between the U.S. Department of Energy, Office of Energy Research, and Oak Ridge Associated Universities.

279

1. INTRODUCTION

Highly cellular soft-tissue tumors which showed no
tendency to calcify sometimes had high concentra-
tions of the isotope [Ga-72]. . . .

G. A. Andrews, S. W. Root, and H. D. Kerman [1]

The above quote, taken intentionally out of context from one of a
series of consecutive papers that appeared in the journal *Radiology*
in 1953, was prophetic, and it is used here for that reason. That
series of articles [2] was, in effect, a negative report on the
clinical use of Ga-72 for diagnosis and radiotherapy of osteogenic
sarcoma and other metastatic cancers of the bone, a possibility
originally suggested by Dudley and his co-workers [3]. In hind-
sight, considering the primitive nature of the nuclear medical
instrumentation that was available at that time and the properties
of Ga-72, this should not have been surprising. That series of
papers also included a preclinical report on the biodistribution of
Ga-67 in rats [4]. Gallium-67 was also briefly studied at our clinic
toward the end of the Ga-72 clinical trials, but no report was made.
Some 15 years later, after it had been recognized that Ga-67 in a
carrier-free form concentrated to a quite high degree in many nonosseous soft-tissue tumors [5,6], a review of the records of the
earlier clinical trials of Ga-67 was made. From this review it
became clear, particularly from the autopsy autoradiographic evidence, that Ga-67 had indeed been rather avidly taken up by some
nonosseous soft-tissue tumors. Apparently, during the studies in
the 1950s, a preoccupation with gallium's uptake at sites of

osteogenic activity (as well as the inadequate instrumentation that
was then available) resulted in a lack of any follow-up of Ga-67's
potential as a tumor-localizing agent. However, as indicated by the
quote above, during the Ga-72 clinical studies in the early 1950s,
it was noted in an incidental way that there were sometimes high concentrations of Ga-72 observed in nonosseous tumor tissues. The concentration that was observed probably occurred only with tracer doses
of Ga-72 when only small amounts of stable gallium would have necessarily been administered. The amount of stable gallium administered
has a pronounced effect on uptake of gallium by tumor tissue, as will
be discussed later.

Because of the chemical similarity of gallium and indium, it is
not unreasonable to speculate that indium radionuclides might show
tumor-localizing properties somewhat similar to those of gallium.
Shortly after the first published report on the avidity of Ga-67 for
tumor tissue, this possibility was confirmed by Hunter and de Kock
[7] using carrier-free In-111. During the intervening years since
the discovery of their tumor-localizing properties, both Ga-67 and
In-111 have been used increasingly for nuclear medical detection of
malignancies, with Ga-67 being used much more so than In-111. More
recently, both radionuclides have also been employed extensively for
the detection of inflammatory processes. Activity-wise the use of
Ga-67 as a radiopharmaceutical is now approximately equal to that of
I-131 [8].

Both gallium and indium have a number of useful radionuclides,
especially those with short half-lives that are available from generators. These short-lived radionuclides are of particular interest
as radiolabels. When used in this manner, these radionuclides take
on the chemical and physical characteristics of the agents they are
bound to rather than acting according to their own elemental natures.
Gallium-68 ($t_{1/2}$ = 68 min) is of particular interest in this respect,
since it is a positron emitter and can therefore be used for positron
tomographic imaging.

In this chapter we review the medical uses of the radionuclides
of gallium and indium, the chemical and biochemical factors involved

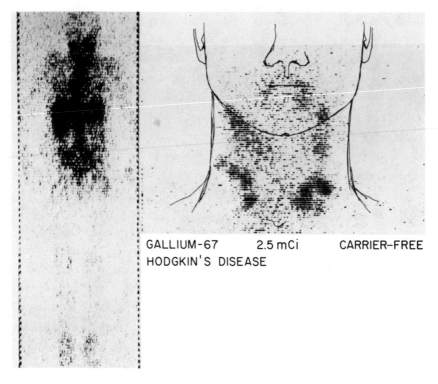

GALLIUM-67   2.5 mCi   CARRIER-FREE
HODGKIN'S DISEASE

FIG. 1. First scan made at Oak Ridge Associated Universities that showed uptake of Ga-67 in the involved lymph nodes of a patient with Hodgkin's disease. (Reproduced with permission from Ref. 5.)

in these uses, and, based on animal studies, the mechanisms that may be involved in their biodistributions. With respect to this latter topic, it is of interest to note that contrary to the usual order in which things occur in the development of pharmaceutical agents, in the case of Ga-67, its potential as a tumor-localizing agent was first observed in humans and then extrapolated to animal models, rather than the reverse. Figure 1 shows a Ga-67 rectilinear scan of the first patient in whom the uptake of Ga-67 in nonosseous tumor tissue was first observed [5].

2. PHYSICAL AND CHEMICAL NATURE
OF GALLIUM AND INDIUM

2.1. Radionuclides

Table 1 shows a tabulation of the pertinent properties of the gallium
and indium radionuclides that have been used medically together with
methods for their production. The fact that Ga-72 could readily be
produced in a reactor was undoubtedly the main reason it was the
first radionuclide of gallium to be investigated clinically. It was,
however, an extremely poor choice, since all the photon emissions of
the radionuclide have energies above 0.6 MeV, with 40% above 1.0 MeV
and 30% above 2.0 MeV. Gallium-67, although far from perfect, does
have acceptable characteristics from a radiation dosimetry and imaging
standpoint. Indium-111 has similar characteristics, but it is superior to Ga-67 from an imaging standpoint. The half-lives of both
Ga-67 and In-111 are quite appropriate, since tumor and abscess imaging with both agents cannot normally be carried out until 1-3 days
have elapsed after they are administered. Short-lived Ga-68, as previously noted, can be used as a radiolabel for positron tomographic
imaging because of its decay mode. More important, it is available

TABLE 1

Characteristics of Some Clinically Useful
Radionuclides of Gallium and Indium

Radionuclide	Half-life	Decay mode	Energy (KeV)	Production
Ga-72	14 hr	β^-	Very high	Reactor
Ga-67	78 hr	E.C.	92, 180, 285	Accelerator
Ga-68	68 min	β^+	511	^{68}Ge decay
In-111	67 hr	E.C.	172, 247	Accelerator
In-113m	1.7 hr	I.T.	392	^{113}Sn decay
In-115m	4.5 hr	I.T.	335	^{115}Cd decay

on demand from a 288-day parent (Ge-68). The original alumina Ga-68
generator devised by Tucker and Greene [9] yields the radionuclide
as the EDTA chelate. Recently, Loc'h and co-workers [10] reported
the development of a new generator with a stannic oxide support that
yields Ga-68 in the chloride form. This should considerably simplify
the preparation of most Ga-68-labeled agents since separation of Ga-68
from EDTA will not be required. The short-lived radionuclide In-113m
has been frequently used in nuclear medical procedures mainly because
of the long half-life of its parent, Sn-113 ($t_{1/2}$ = 115 days), although
In-113m's photon energy is somewhat high for scintillation imaging.
Indium-115m has been used infrequently, mainly because its parent,
Cd-115, has only a 54-hr half-life.

2.2. Relationship to Other Elements

Gallium and indium fall into group III in the periodic table under
aluminum. Their trivalent forms are the most stable valence state
and, like aluminum, they are both amphoteric. Neither gallium nor
indium have any known biological trace element function; in fact,
indium is quite toxic while gallium is only moderately so. From a
purely chemical standpoint, it does seem rather remarkable that such
simple agents--simple compared to most bioactive substances--should
show the affinities for tumor tissue that they do.

One factor of possible importance in the biodistribution of
Ga(III) may well be its ionic radius (0.62 Å). Iron(III), which is
carried in the plasma bound to the protein transferrin, has a very
similar ionic radius (0.64 Å). Gallium(III) also binds to transferrin, presumably at the same or similar sites that Fe(III) does
[11,12]. This binding of gallium to transferrin has been strongly
implicated in both the initial and final biodistribution of gallium
[13,14]. Both In(III) (ionic radius, 0.81 Å) and Sc(III) (ionic
radius, 0.81 Å) also bind to transferrin and even more strongly than
does Ga(III) [15-17]. (The relevance of Sc(III) to the biodistribution of Ga(III) will be discussed in Sec. 4.1.4.)

2.3. Ion Hydrolysis

Both gallium and indium, like aluminum, are amphoteric, gallium more
so than either indium or aluminum. The resultant hydrolysis of the
trivalent cations of these two elements would appear to be an important factor that may control their uptakes in normal and abnormal
tissues following parenteral administration of these substances.
[Neither Ga(III) nor In(III) is absorbed to any appreciable extent
after oral administration.] Although both Ga(III) and In(III)
initially bind strongly to plasma transferrin [In(III) more so than
Ga(III)], there are necessarily equilibria that are established
between the plasma protein-bound and free forms of these elements.
The natures of the free forms are in turn dictated by the hydrolytic
processes which take place at the hydrogen ion concentration of the
environment in which the free forms exist. In the case of Ga(III),
the hydrolytic process proceeds in the following manner:

$$Ga^{3+} + H_2O \rightleftharpoons Ga(OH)^{2+} + H^+ \tag{1}$$

$$Ga(OH)^{2+} + H_2O \rightleftharpoons Ga(OH)_2^+ + H^+ \tag{2}$$

$$Ga(OH)_2^+ + H_2O \rightleftharpoons Ga(OH)_3 + H^+ \tag{3}$$

$$Ga(OH)_3 + H_2O \rightleftharpoons Ga(OH)_4^- + H^+ \tag{4}$$

The hydrolysis of In(III) proceeds in a similar way. Figure 2 shows
the distribution of the various mononuclear forms for both elements
as a function of pH as published by Baes and Mesmer [18]. Although
polynuclear hydrolytic forms for both elements do appear to occur
[as is shown in Fig. 2 for In(III)], at the low molar concentrations
involved with the carrier-free amounts of Ga-67 and In-111 that are
used in nuclear medical procedures, such forms are undoubtedly of
little importance. From Fig. 2 it is apparent that both Ga(III) and
In(III) will be present to a large extent as the neutral hydroxides
at physiological pH. This fact and the changes in hydrolytic form
that would result from a lowering of the pH may well be of major
importance in the biodistribution of Ga-67 and In-111. Note in
Fig. 2 that, as stated at the beginning of this section, gallium is

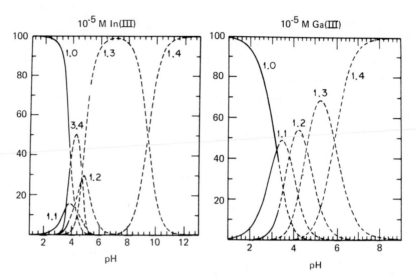

FIG. 2.   Hydrolysis of Ga(III) and In(III) as a function of pH [18].
Notations on curves indicate number of OH⁻ groups associated with the
metal ions, e.g., 1,3:M(OH)$_3$.   (Reproduced from Ref. 18 by permission
of John Wiley & Sons, Inc.)

considerably more amphoteric than indium.   In fact, after precipitation gallium hydroxide is quite soluble in excess base, whereas
indium hydroxide is only slightly so [18].

3.   CLINICAL USES

3.1.   Malignancies

The historical background behind the present use of Ga-67 for the
detection of neoplasms has been treated in some detail elsewhere
[19].   Indium-111 has been used far less frequently as a tumor-
localizing agent than Ga-67.   A number of factors are involved in
the affinities that these two radionuclides have for malignant
tissue, e.g., carrier effects and binding by transferrin.   These
factors will be discussed in Sec. 4.   Interestingly enough, stable
gallium in the nitrate form is at present being tested in clinical
trials as a chemotherapeutic agent for cancer.

3.1.1. *Cancer Detection*

The blood clearance of intravenously administered Ga-67 citrate and
In-111 chloride (the forms usually employed) is normally rather slow
and 1-3 days is usually required before optimum imaging of these
agents can be achieved. There tends to be a rather high uptake of
Ga-67 and In-111 by normal liver, more so for the latter. Both
agents deposit to some extent in the bone and, in the case of In-111,
also in the bone marrow. Concentration of Ga-67 in the fecal matter
in the lower portions of the intestinal tract frequently constitutes
a problem in scan interpretation with Ga-67, and repeat scans are
then required. This does not appear to occur as often with In-111
[20].

The medical literature now abounds in reports on the use of
Ga-67 and In-111 to detect many different types of malignancies in
humans [21,22]. Table 2 lists some of these together with selected
references. By 1972 it was apparent that lung cancer, Hodgkin's
disease, and other lymphomas could be quite effectively detected
with Ga-67 [33]. It was shown that malignant processes in the liver
could also be detected as areas of increased Ga-67 uptake, despite
the avidity that normal liver has for Ga-67 [34]. More recently,

TABLE 2

Malignancies in Humans That Have Been
Detected by Ga-67 and In-111 Imaging

Type	Ref.
Lung cancer	[23,24]
Hodgkin's disease	[25]
Non-Hodgkin's lymphoma	[26-28]
Malignant melanoma	[29]
Breast carcinoma	[30]
Ewing's sarcoma	[31]
Tumors of the genitourinary tract	[32]

Ga-67 has been used effectively in the management of malignant melanoma [35].

Although there are still some proponents of the use of In-111 in cancer diagnosis [20], the consensus now seems to be that Ga-67 is superior to In-111 for that purpose. Following the finding that Co-57 bleomycin had tumor-localizing properties, it was proposed that the use of bleomycin might also improve the diagnostic effectiveness of In-111 [36]. This has not been found to be the case by most investigators, however. Our own animal studies have indicated that there is little to be derived from the use of bleomycin with In-111. Table 3 shows the results of one such study. This study suggests that, although the tumor-to-nontumor ratios of several tissues were increased, others were decreased and the high loss of In-111 through excretion probably more than offsets any benefits gained by using bleomycin.

Although neoplastic processes not found with other techniques are occasionally detected with Ga-67, it has now become clear that

TABLE 3

Comparison of the Tissue Distribution of In-111 Citrate
and In-111 Bleomycin in Rats Bearing RFT Tumors

	4 hr		24 hr	
	Citrate	Bleomycin	Citrate	Bleomycin
Tumor conc. (%/g)	0.34	0.15	0.61	0.20
Ratio tumor conc. to:				
Liver	0.18	0.07	0.23	0.08
Kidney	0.10	0.10	0.14	0.12
Lung	0.07	0.34	0.23	1.30
Muscle	2.70	4.60	4.20	7.10
Marrow	0.15	0.18	0.20	0.19
Blood	0.12	0.50	0.83	1.70
Percent excretion	4.0	64.0	6.8	65.0

Source: R. L. Hayes et al., unpublished results.

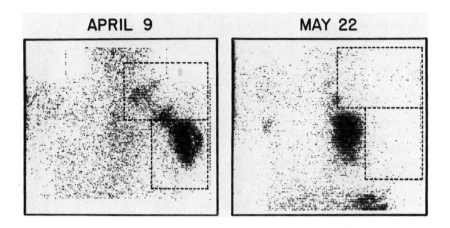

FIG. 3. Followup Ga-67 scan (right) of a patient given Co-60 teletherapy for treatment of reticulum cell sarcoma (scan before treatment on left). Note recurrence of disease outside Co-60 treatment
fields (dashed lines) in scan on right.

the main benefits to be derived from the use of Ga-67 (and, for that
matter, In-111) in cancer diagnosis are in staging and follow-up.
Figure 3 shows an example of a Ga-67 follow-up study of a patient
who had had Co-60 teletherapy for treatment of reticulum cell sarcoma.

3.1.2. Cancer Chemotherapy

It came as a surprise (certainly to these observers) that, in addition
to having diagnostic utility in a radionuclide form, gallium as the
stable element also appeared to have some possible application as a
cancer chemotherapeutic agent. This finding resulted from studies
carried out at the National Cancer Institute at Bethesda [37]. As a
result, investigational new drug trials of gallium nitrate have been
carried out at a number of medical institutions in the United States.
To date the results have not been too promising.

3.2. Inflammatory Processes

During the early stages of our clinical investigation of Ga-67 as
a tumor-localizing agent, it was recognized that Ga-67 tends to

TABLE 4

Inflammatory Processes in Humans That Have Been
Detected by Ga-67 and In-111 Imaging

Type	Ref.
Abscesses	[41]
Sarcoidosis	[42]
Pneumocystis carinii	[43]
Bleomycin toxicity	[44]
Bubonic plague	[45]

concentrate in inflammatory processes such as healing areas following
surgery. It was thought at that time that this fact might limit the
usefulness of Ga-67 for the detection of malignancies, if, in general,
inflammatory lesions showed enhanced uptakes of Ga-67 [6]. This has
not, however, been the case; in fact, Ga-67 is probably now used to
a greater extent for the detection of inflammatory processes, such as
postsurgical abdominal abscesses, than it is for the detection of
cancer [38]. To detect these types of lesions, Ga-67 is generally
administered intravenously in the citrate form. Indium-111 chloride
also concentrates in inflammatory processes [39] and is accordingly
used for the detection of such lesions, but it is now normally given
in the form of In-111-labeled autologous white blood cells [40]. The
labeling process used is discussed in Sec. 3.3.1. Table 4 contains
a listing of some of the types of inflammatory processes that have
been detected through the use of Ga-67 and In-111. Indium-111-
labeled white blood cells appear to be more effective for the detection of acute inflammatory processes, while Ga-67 citrate tends to
localize to a much greater extent in chronic inflammatory lesions.
This may possibly be due to the avidity that Ga-67 seems to have for
macrophages. It would appear, judging from the number of chronic
inflammation-associated medical problems listed in Table 4, that
Ga-67 is probably attracted to any site where there is, for whatever
reason, an increased concentration of macrophages.

3.3. Use as Radiolabels

As previously mentioned, when Ga-67(III) and In-111(III) are administered parenterally in a readily dissociable form, they bind more
or less immediately to plasma transferrin. Further biodistribution
processes are then mainly governed by this binding to transferrin.
If these two radionuclides are, on the other hand, administered in
forms that do not appreciably dissociate, obviously little binding
to transferrin occurs, and Ga-67 and In-111 will then take on the
biodistribution characteristics of the nontransferrin substance to
which they are in turn bound. They will then be acting, from an
imaging standpoint, as radiolabels.

3.3.1. *Blood Elements*

Both gallium and indium radionuclides have been used to label white
blood cells, erythrocytes, and platelets [40]. These radiolabeled
agents have been used to image inflammatory lesions, the blood pool,
and thrombotic processes. In vitro labeling of these different cell
types with Ga-67 and Ga-68, following cell separation, is not very
efficient and, in the case of white blood cells, in vivo labeling
is normally used instead because no particular advantage seems to
be gained by using in vitro labeling techniques [46]. With In-111,
on the other hand, in vitro techniques are quite effective, particularly when the 8-hydroxyquinoline (oxine) chelates of In-111 are
used [40]. An In-111 oxine preparation that has been approved by
the U.S. Food and Drug Administration is now being supplied commercially.

3.3.2. *Chelates and Chelate Derivatives*

Since, as mentioned above, strong chelating agents such as ethylenediaminetetraacetic acid (EDTA) form complexes with polyvalent ions
that are quite stable, it is to be expected that gallium and indium
chelates with such substances will behave, because of their low dis-
sociabilities, in entirely different ways than they do in their free

or easily dissociable states. Thus chelates of gallium and indium
radionuclides, such as those with EDTA, are rapidly cleared from the
blood and excreted in the urine. Chelates of these radionuclides
thus behave like the chelators themselves.

In addition to forcing gallium and indium radionuclides to
behave biodistribution-wise in a manner similar to the substances
to which they are chelated, it is also possible to go one step further and in turn use radiochelates as labels themselves. Thus, if
chelate derivatives of proteins are prepared, these can in turn be
used, when labeled with gallium and indium radionuclides (or, for
that matter, any other suitable radionuclide), to follow the biodistribution of selected chelate derivatives of proteins. Gallium-67-
and In-111-labeled DTPA chelate derivatives of albumin and fibrinogen
are two such examples [47,48]. This approach to the design of radiopharmaceuticals is most intriguing when one considers the possibilities that present themselves with the radiolabeling of chelate
derivatives of monoclonal antibodies, particularly antibodies to
specific tumor types in humans.

3.3.3. Macroaggregates and Colloids

As pointed out in the foregoing, gallium and indium radionuclides
can be effectively used as radiolabels for simple organic molecules
having specific biodistribution characteristics and also as radiolabels for more complex macromolecules, such as proteins. It is also
possible in turn to label certain colloids and macroaggregates with
these radionuclides for use in imaging of the lung, liver, spleen,
and bone marrow. Advantage can be taken of the affinity that gallium
and indium radionuclides have for denatured macroaggregated and micro-
sphere forms of human serum albumin [49-51]. (Macroaggregates in the
range of 20-100 µm are physically trapped in the capillary bed in the
lungs.) The exact reason for this binding is unclear. Hnatowich and
co-workers [52] also reported the Ga-67 labeling of DTPA chelate
derivatives of albumin microspheres, in which case the reason for
labeling is quite clear. Alizarin [53] and hydrous ferric oxide

preparations [54,55] have been used as carriers for gallium and
indium radionuclides for imaging the lung and the reticuloendothelial
system. With the latter, the technique involves the heat hydrolysis
of dilute solutions of ferric chloride in the presence of Ga-68. By
varying the ferric chloride concentration and the heating rate, the
size of the colloid particles can be varied over a wide range [55].
Use of even smaller colloid particles ($\sim$2 nm) increases the uptake
of the radiocolloid in the bone marrow. The pH of the solution
before hydrous ferric oxide colloid formation is initiated is of
considerable importance [56]. If the pH is too low, Ga-68 labeling
of the colloid is impaired. Indium radionuclides cannot be used with
this type of labeling technique because indium is less amphoteric
than gallium (see Sec. 2.3).

4. FACTORS INVOLVED IN TISSUE UPTAKE

As previously indicated, many factors enter into the biodistributions
of both gallium and indium radionuclides. To the authors, these
effects are indeed quite fascinating because they do seem to provide
some basic clues to the mechanisms involved in the uptake of Ga-67
and In-111 by malignant tissues and inflammatory processes. An understanding of the mechanisms involved could thus lead to the identification of new and better radiopharmaceutical agents for detecting such
diseases. Present knowledge of the factors involved in the biodistribution of gallium and indium radionuclides come mainly from studies
in animals. In tumor studies, animals bearing transplanted tumors
have been quite valuable. The 5123C Morris hepatoma in Buffalo rats
has the highest relative Ga-67 uptake of all of the rodent tumor
systems we have investigated [33].

4.1. Gross Tissue Distributions

As was pointed out in Sec. 3.1, gallium and indium radionuclides
tend to concentrate mainly in liver, bone, and, to some extent,

bone marrow. It has been found that the viable portions of tumor tissues invariably have a higher concentration of Ga-67 than necrotic portions [33]. In general, the same also appears to occur with In-111.

4.1.1. Effect of Chemical Form

Although with an increase in dose there would necessarily be no change in the forms in which gallium and indium radionuclides are administered, nevertheless the amount of the stable elements (mg/kg) that are administered can be extremely important as the dose is increased. This will be discussed in detail in Sec. 4.1.4.a.

As was pointed out in Sec. 3.3.2, the administration of a non-dissociable form of gallium or indium can lead to entirely different biodistribution patterns. Accordingly, it is tempting to assume that the higher the stability constants of the chelates of gallium and indium, the more the gallium or indium label will behave in vivo like the unbound chelating ligand itself. This is, however, not necessarily the case. For example, we have found in aminal studies that although the stability constant of the gallium chelate of diethylene-triaminepentaacetic acid (DTPA) is $\sim 10^4$ higher than that of gallium EDTA, the biodistribution of Ga-67 DTPA is actually similar to that of low-stability-constant Ga-67 citrate, rather than that character-istic of Ga-67 EDTA [55] (see Fig. 4). This apparent contradiction may possibly be explained by the fact that Ga-67 EDTA dissociates at an extremely slow rate ($t_{1/2}$ = ~ 20 days) at pH values in the physio-logical range [57]. Apparently Ga-67 DTPA, although it has a much higher stability constant than Ga-67 EDTA, dissociates readily at physiological pH and consequently is "pulled apart" by the mass action effect produced by the level of transferrin that is present in the plasma, while Ga-67 EDTA is not, in spite of its lower sta-bility constant. A similar but reverse situation apparently also occurs with indium, i.e., In-111 DTPA retains its integrity in vivo, although the stability constant of this chelate is far less than that of In-111 transferrin [15]. Apparently a very slow rate of dissociation of In-111 DTPA chelate is involved. Although these

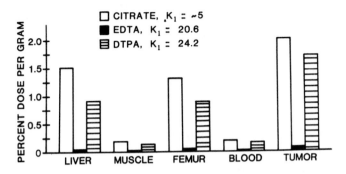

FIG. 4. Effect of different chelating agents on the 24-hr tissue distribution of Ga-67 in tumor-bearing rats.

two isolated peculiarities are undoubtedly somewhat unique, the fact that they do occur raises the question as to whether chelates of gallium and indium radionuclides that have rather low stability constants might actually be quite useful, if only their rates of dissociation were sufficiently slow [55].

With respect to form, when the imaging of malignant and inflammatory lesions with Ga-67 are involved, it seems to make little difference as to what the counter ion(s) is [55]. The counter ion should however be one that will insure against the hydrolysis of gallium at the pH to which the Ga-67 preparation has been adjusted, since otherwise the small amounts of Fe(III) that are invariably present in such preparations will result in a Ga-67-labeled iron oxide colloid, unless the pH of the preparation is quite low (see Sec. 3.3.3). With In-111 this generally does not constitute a problem, although an intentional adjustment of the pH can make an appreciable difference in its biodistribution [58].

4.1.2. Effect of Sex, Age, and Food Consumption

In the early stages of our investigation of the biodistribution of Ga-67 in rodents it became evident that males and females have, in several respects, quite different uptakes of Ga-67 in normal tissues [59]. Accordingly, we exercised control over this factor in our further studies. (The Ga-67 tumor-to-muscle ratio in female rats is four to five times higher than it is in males; no difference in percent per gram uptake of Ga-67 in tumor tissues has been observed, however.) Possibly these effects occur with In-111 also.

Age (in rodents) also appears to have an effect on the biodistribution of Ga-67 [59]. An increase in the uptake of Ga-67 by the reticuloendothelial (RE) tissues occurs. No similar effect has been observed in humans [60]. Food consumption seems, likewise, to be of importance [61]; fasting produces a pronounced increase in the concentration of Ga-67 in the RE system.

4.1.3. Effect of Pregnancy and Lactation

Concentration of Ga-67 in the mammary glands of lactating females was first reported by Larson and Schall [62] and later confirmed by Fogh [63] in clinical studies. This can possibly be explained by the increase in lysosomal enzymes that occurs in mammary tissue during and after pregnancy [64]. Gallium-67 also passes through the placenta of hamsters and is taken up by the fetus [65]. During pregnancy the uptake of Ga-67 in the mammary glands also increases dramatically, and following birth the Ga-67 contained in the dam's milk is for a time freely absorbed from the intestinal tract of the nursing infant [65].

4.1.4. Effect of Altered Binding to Plasma Proteins

Hartman and Hayes [66], using Ga-72 and a dialysis technique, were the first to observe that gallium was bound to plasma proteins and that the proteins mainly responsible were transferrin and, to a smaller extent, ceruloplasmin [67]. That gallium is bound by transferrin is now generally accepted, and this fact has been used by several groups as the basis for an explanation of the mechanism involved in the preferential uptake of Ga-67 by malignant tissues, although some of these proposed mechanisms are quite at odds with each other. Many suggestions as to the mechanism involved with inflammatory processes have been made. It seems quite possible that indium uptake by tumors and inflammatory lesions is similar to that occurring with gallium (see Secs. 4.2.1 and 4.2.2).

   4.1.4.a. The Stable Elements. It was first observed during the early 1950s, while clinical trials of Ga-72 were underway, that the amount of stable gallium that was administered with Ga-67 had a pronounced effect on the tissue distribution and excretion pattern [4]. How dramatic that effect was did not become apparent until 1965 when it was shown that the addition of stable gallium to Ga-68 ($t_{1/2}$ = 68 min) produced a rapid and greatly enhanced deposition of that radionuclide in bone [68] (see Fig. 5). As a result, until the advent of Tc-99m phosphonates for bone scanning, Ga-68 with added stable gallium was investigated as an agent for bone imaging in our clinic.

   The reason for the effect that stable gallium has on the bio-distribution of gallium radionuclides became apparent from an in

⁶⁸Ga CITRATE

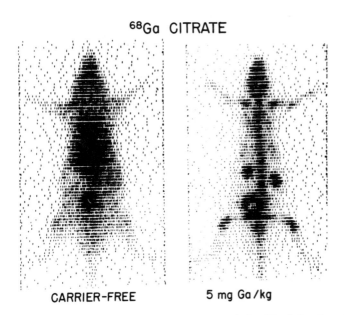

CARRIER-FREE   5 mg Ga/kg

FIG. 5.   One-hour scans of rats administered Ga-68 without and with added stable gallium.   (Reproduced with permission from Ref. 68.)

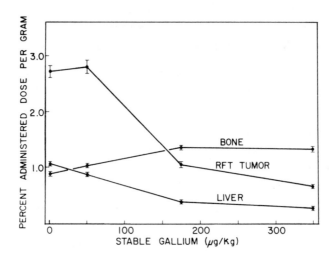

FIG. 6. One-day tissue distributions of Ga-67 in tumor-bearing rats receiving increasing amounts of stable gallium. (Reproduced from Ref. 19 by permission of Grune & Stratton, Inc.)

vitro study which showed that the binding of gallium to plasma proteins was saturable and thus that at stable gallium levels of ⌁5 mg/kg, there was no significant binding of gallium to plasma proteins [66], in keeping with the observed changes seen in the early biodistribution of Ga-68 (Fig. 5). It was later found, after Ga-67 was recognized as an effective nonosseous tumor-localizing agent, that tumor tissue binding of Ga-67 was also saturable by stable gallium to approximately the same extent as normal soft tissues [19] (see Fig. 6). From Fig. 7, where the results of a study of the effect of the level of administered gallium on In-114m are shown, it would appear that a similar binding site saturation process is also involved with indium.

4.1.4.b. *Iron Status.* Since Ga(III) binds to plasma transferrin, and probably at the same binding sites where Fe(III) does [11], it is not unreasonable to assume that changes in plasma iron status could have pronounced effects on the plasma protein binding of Ga-67 and the consequent biodistribution of this agent. Such is indeed the case. This was shown in an early animal study by Higasi

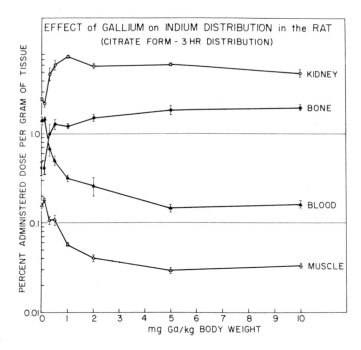

FIG. 7. Effect of stable gallium on the tissue distribution of
high-specific-activity In-114m in rats. (From R. L. Hayes et al.,
unpublished results).

and co-workers using an iron oxide colloid preparation [69]. More
recently this iron saturation technique, with consequent decreased
binding of Ga-67 to plasma proteins, has been used clinically with
modest success to achieve increased tumor-to-nontumor ratios [70].
Our own studies with both iron colloid and ferric citrate have confirmed this effect [13]. More important, in these studies we also
observed that, while the Ga-67 uptake in normal tissues was decreased
as expected, the uptake in tumor tissue remained the same or increased. Hyperferremia has been implicated as the major factor
involved in the reduction in tissue uptakes and the consequent
lowered whole-body retention of Ga-67 that is observed following
whole-body x-irradiation [71], again stressing the importance of
the binding of Ga-67 to plasma transferrin. The importance of iron

status is further emphasized by the work of several groups who found
that deferoxamine, a pharmaceutical used to treat iron overload, was
effective in decreasing the uptake of Ga-67 in normal tissues [72].

4.1.4.c. *Scandium Administration.* During the period in which
we were concerned with attempting to find an agent other than stable
gallium to block the binding of Ga-68 to plasma proteins and thus
promote the deposition of this radionuclide in bone, we tested a
number of other trivalent elements that had ionic radii and chemical
natures similar to that of gallium. To our pleasant surprise, we
found that Sc(III) not only had the desired effect but also produced
this effect at a much lower dose level than that required with stable
gallium. The possibility of using scandium to enhance Ga-68 bone
deposition fell by the wayside when Tc-99m phosphonates were introduced for bone scanning. However, after the preferential uptake of
Ga-67 by nonosseous tumor tissue was recognized, we did, for obvious
reasons, test the effect of stable Sc(III) on the uptake of Ga-67 by
transplanted rodent tumors. To our delight we found that not only
did the uptake of Ga-67 decrease in normal tissues (other than bone),
but the concentration in tumor tissues remained the same or in some
cases increased [73].

Similar effects, but less dramatic, have been reported by
another group [74]. Thus Sc(III) probably blocks Ga-67 binding sites
on transferrin, but does not block the binding sites for Ga-67 in
tumor tissue. (Scandium, in subsequent approved clinical trials, was
found to be toxic to humans, although toxicological studies in animals
had indicated that this would not be the case [75].)

4.1.4.d. *Plasma Transferrin Level.* The fact that iron and
scandium administrations decrease the uptake of Ga-67 in normal soft
tissue while the uptake in tumor tissue remains approximately the
same (marginal for iron, dramatic for scandium) indicated to us that
there were probably basically different pathways involved in the uptake of Ga-67 by normal and malignant tissues. We proposed such in
a recent paper [13] (see Fig. 8). Since this suggestion on our part
necessarily implied that an increase in plasma protein binding of

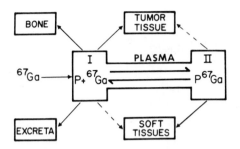

FIG. 8.   Schematic of possible pathways that Ga-67 may take in its biodistribution in normal and tumor tissue [13].   P stands for protein.   (Reproduced with permission from Ref. 13.)

Ga-67 would, in turn, increase the uptake of Ga-67 in normal tissues and at the same time decrease the uptake in tumor tissues (again see Fig. 8), we decided to carry out appropriate tests of this hypothesis by increasing the transferrin level through both endogenous and exogenous means.   The results were completely in line with the pathways suggested in Fig. 8, i.e., a decrease in tumor uptake of Ga-67 was observed, while at the same time the uptake in normal tissues was increased [13].

4.2.   Subcellular Distributions

4.2.1.   At the Organelle Level

It is now generally agreed that at the subcellular level, Ga-67 is mainly associated with lysosomal structures.   Swartzendruber and his associates, who used electron microscopic autoradiography, were the first to make this observation [76].   Figure 9 is taken from their original report.   Brown and co-workers confirmed this observation using ultracentrifugal fractionation techniques [77].   In a continuing more definitive study, they also found that subcellular Ga-67 was probably more closely associated with a smaller lysosomal-like particle that resided in the endoplasmic reticulum vesicle fractionation region, and further that this Ga-67-associated organelle was more prevalent in tumor tissue than it was in normal tissue

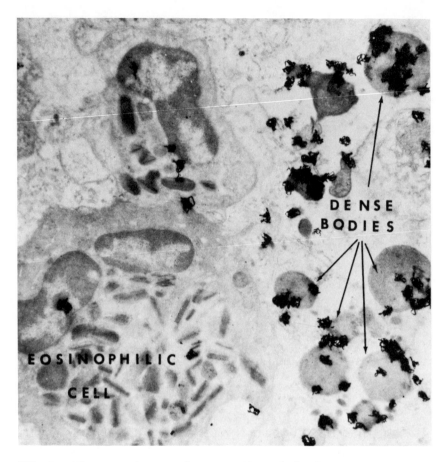

FIG. 9. Electron microscopic autoradiogram showing the concentration of Ga-67 in lysosomes as indicated by the number of silver grains on and near "dense bodies." (Reproduced with permission from Ref. 76.)

[78] (see Fig. 10). This latter study resulted from an improvement in the ultracentrifugal fractionation technique for separation of tumor tissue organelles. The new technique involved the removal of the mucilaginous material that appears to be generally present at high levels in tumor tissue. The presence of this mucilaginous material apparently causes smaller subcellular organelle particles to be spun down in the nuclei fraction. This fact could readily explain the reports that have appeared in the literature [79,80],

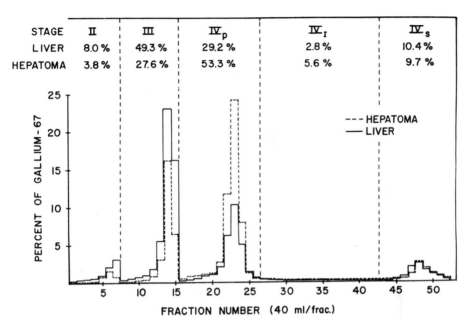

FIG. 10. The distribution of Ga-67 between the subcellular particles (excluding nuclei) found in normal liver and 5123C Morris hepatoma [78]. The roman numerals indicate: II, mitochondria; III, lysosomes; IV_p, endoplasmic reticulum vesicles; IV_I, intermediate fraction between IV_p, and IV_s; IV_s, soluble fraction. (Reproduced with permission from Ref. 78.)

which seem to indicate that Ga-67 is associated to a great extent with nuclei. With this new fractionation technique, only about 5% of the Ga-67 present in tumor tissue was found to be associated with cell nuclei, in agreement with autoradiographic observations [76]. Indium-111 has, not surprisingly, also been found to show a subcellular organelle association pattern similar to that of Ga-67 [39].

4.2.2. At the Macromolecular Level

This is perhaps the most intriguing part of the gallium and indium story, i.e., the difference between the association of gallium and indium in tumor and normal tissues is indeed quite unique at the macromolecular component level. We have found in our work at this

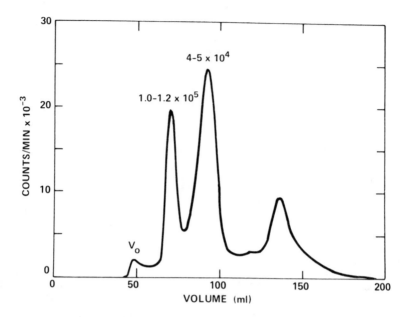

FIG. 11.   Gallium-67 Sephadex G-200 gel filtration profile of a distilled water extract of 5123C Morris hepatoma tissue. (Reproduced with permission from Ref. 81.)

laboratory with aqueous extracts of tumor and normal tissues that the binding of Ga-67 to an ~45,000-dalton (40-K) glycoprotein (see Fig. 11 [81]) is much higher in tumor tissue than in normal tissues (~50% vs. ~10%) [82]. Furthermore, when the much higher uptake of Ga-67 by tumor tissue, compared with that seen in normal tissue, is taken into account, it appears that the association of Ga-67 with this 40-K protein in tumor tissue is 9-65 times that of the association that occurs in normal tissues [83]. Perhaps not surprisingly, In-111 also appears to behave in a similar manner [82] (see Fig. 12). The 40-K Ga-67-binding protein is heat- and pH-sensitive and can be saturated with stable gallium by the administration of as little as 25 μg of stable gallium per kg of body weight [82].

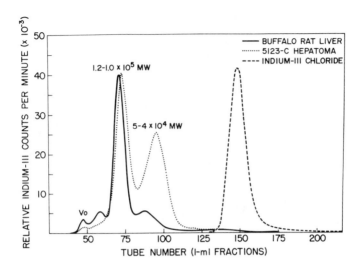

FIG. 12.   Indium-111 Sephadex G-200 gel filtration of distilled water extracts of normal liver and 5123C Morris hepatoma tissues. (Reproduced with permission from Ref. 82.)

5.   MECHANISMS

It is not only of academic interest to determine what mechanism(s) and processes are involved in the uptake of gallium and indium radionuclides by malignant and inflammatory lesions, but more important is the fact that an understanding of these processes may lead to the identification and design of new and better radiopharmaceuticals.

5.1.   Interactions with Transferrin, Lactoferrin, Siderophores, and Ferritin

As has previously been pointed out, the binding of Ga-67 (and undoubtedly that of In-111) to transferrin is of prime importance in the initial and final biodistribution of these two radionuclides. Based mainly on in vitro studies, it has been proposed that transferrin receptor sites on tumor cells, to which the Ga-67 transferrin

TRANSFERRIN RECEPTOR HYPOTHESIS

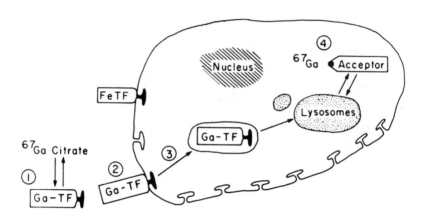

Step 1   Transferrin binding
Step 2   Receptor binding
Step 3   Internalization and binding to lysosomes
Step 4   Binding to "acceptor" macromolecules

FIG. 13. Schematic of the pathway proposed by Larson and co-workers
[14] for the uptake of Ga-67 by tumor tissue. (Reproduced with permission from Ref. 14.)

complex binds, are primarily involved in the enhanced uptake of Ga-67
by malignancies. Larson and his co-workers [14] suggested the pathways shown in Fig. 13, and have pointed out the similarity of the
binding of gallium and iron to transferrin as has the Sephton group
[84]. Sephton and his co-workers have, however, recently stressed
the importance of extravascular transferrin, since they now appear
to feel that the apparent correspondence, in some respects, between
gallium and iron metabolism is misleading [85]. Actually, the con-
clusions drawn from in vitro Ga-67 uptake studies appear to be somewhat suspect since there seems to be no similarity at all in the
subcellular distributions of Ga-67 in tumor cells (5123C Morris
hepatoma) that have been labeled by both in vivo and in vitro techniques [86]. That Ga-67 and In-111 binding to plasma proteins is
of importance is certainly not to be contested, however, especially

since we ourselves feel that the binding of Ga-67 by transferrin is a key factor in the biodistribution of gallium radionuclides (again see Fig. 8).

It has been pointed out by Hoffer [87] that both lactoferrin and bacteria-derived siderophores could account for at least a part, if not all, of the Ga-67 uptake that is observed with inflammatory processes. The uptake of Ga-67 by microorganisms themselves has also been implicated [88]. The assertion that the binding of Ga-67 to the ferritin that is present in tumor cells could account by itself for the unusual affinity that Ga-67 has for malignant tissue does not seem to these reviewers to be too well supported.

5.2. Interchange at Calcium Binding Sites

Since gallium radionuclides, under proper conditions, appear to behave like the element calcium with respect to bone deposition, it is not unreasonable to suspect that uptake of Ga-67 by tumor cells might be related in part to the alkaline earth elements. This view has been promoted by Anghileri and co-workers [89]; however, their evidence has not as yet received wide acceptance.

5.3. Increase in Blood Flow and Capillary
Bed Permeability

By whatever route a substance is administered, its localization at sites other than the point of entry will usually involve transport to such sites by the blood. This necessarily raises the question of the degree of vascularity as well as the rate of blood flow in such localization sites. The degree to which the administered sub-stance penetrates the capillary bed and the size of the interstitial fluid pool obviously will be important factors in its tissue uptake. It has been suggested by Potchen and co-workers [90] that these factors may be involved in the preferential uptake of Ga-67 by

malignancies. Tzen et al. [91] also suggested that alterations of
these processes from normal may be significant in the uptake of
Ga-67 by inflammatory processes. However, in view of other basic
research involving blood flow measurements, it does not seem that
this would be the case [92], although the other work did indicate
that the interstitial fluid volumes in tumors (particularly that in
the 5123C Morris hepatoma) were much higher than that found in normal
tissues.

5.4. Variations in Tumor and Normal Tissue pH

The existence of a decreased hydrogen ion concentration in tumor
tissue due to anaerobic glycolysis has been suggested as one of the
factors that may be involved in the preferential uptake of Ga-67 by
malignancies [93]. The pH of the internal millieu of experimental
animal tumors is reported to be 0.3-0.4 of a pH unit below that of
normal tissue [92]. The effect that this reduction in pH might have
on the relationship between the various hydrolytic forms of gallium
can be seen in Fig. 2. Although the effect brought about by decreased
pH would be rather small, nevertheless the concentration of neutral
$Ga(OH)_3$ should be increased in tumor tissue. As is pointed out in
Sec. 5.5, the uncharged hydroxide form of gallium may be of major
importance in the initial uptake of Ga-67 by passive diffusion. If
so, an increase in the concentration of that form by a still further
decrease in pH should in turn increase Ga-67 tumor uptake. Such an
increase in Ga-67 uptake by experimental tumor tissues has been
demonstrated recently by Vallabhajosula and co-workers in experiments
designed to decrease tumor tissue pH using glucose [94].

5.5. Plasma Membrane Permeability and Endocytosis

Hyperpermeability of the plasma membrane of tumor cells has been
proposed by our group as one of the main factors involved in the

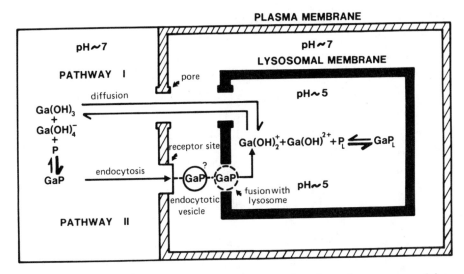

FIG. 14. Schematic of two basic pathways that have been proposed by Hayes and co-workers [96] for the entry of Ga-67 into soft-tissue tumors, normal soft tissue, and inflammatory processes. It is proposed that uptake of Ga-67 by tumor cells proceeds by pathway I and that entry into normal tissue and inflammatory processes proceeds by pathway II. (Reproduced with permission from Ref. 96.)

uptake of Ga-67 by malignant tissues [13]. Anghileri [95] and others have also speculated that hyperpermeability may be an important factor. This could in turn be the case with In-111. From our animal studies [13] (see also Sec. 4.1.4), we feel that, because of its amphoteric nature (see Sec. 2.3), gallium probably enters tumor cells in a neutral form [$Ga(OH)_3$] by passive diffusion and then, in turn, in the same form, penetrates the membranes of the lysosomal organelles that are present in the cytosol. Once in the lysosomes, the low intralysosomal pH (4-5) forces gallium into cationic forms [mainly $Ga(OH)^{2+}$ and $Ga(OH)_2^+$; see Fig. 2], which in turn bind more or less irreversibly to intralysosomal proteins. This process is indicated in Fig. 14 as pathway I. Although Ga-67 could also enter normal cells by pathway I (Fig. 14), the experimental evidence we obtained in animals [13,73] indicates that the main pathway involved in the entry of Ga-67 into normal soft tissues is not pathway I, but rather probably an endocytotic process (pathway II in Fig. 14). Based on further work in animals bearing an experimental abscess [96], we have concluded that the pathway for Ga-67 entry into inflammatory processes is probably more closely related to that involved with normal soft tissues (pathway II, Fig. 14) than it is with that of malignant tissues (pathway I), and that possibly the increased macrophage population that is present in such lesions is responsible for their Ga-67 avidity.

6. SUMMARY AND CONCLUSIONS

The radionuclides of gallium and indium have unique and useful applications in nuclear medicine. Not only do they exhibit unusual affinities for specific types of lesions, i.e., malignancies and inflammatory processes, but they can also be used very effectively as radiolabels for many radiopharmaceutical agents that in turn can be used to monitor the functional status of normal organs and diseases of such organs. It is probable that in the future use of gallium and possibly indium radionuclides will increase, and medical science and those being served by it will benefit accordingly.

Although it appears that trivalent gallium and indium radionuclides are somewhat unique because of their avidities for malignant and inflammatory processes, an understanding of why this happens might very well give rise to the design and development of an entirely new and wide-ranging class of radiolabeled organics that in turn will have far better lesion-localizing characteristics than those of either gallium or indium. Hopefully such will occur!

ABBREVIATIONS

EDTA   ethylenediaminetetraacetic acid
DTPA   diethylenetriaminepentaacetic acid

REFERENCES

1. G. A. Andrews, S. W. Root, and H. D. Kerman, *Radiology, 61,* 570 (1953).

2. M. Brucer et al., *Radiology, 61,* 534-613 (1953).

3. H. C. Dudley, G. W. Imirie, Jr., and J. T. Istock, *Radiology, 55,* 571 (1950).

4. H. D. Bruner, R. L. Hayes, and J. D. Perkinson, Jr., *Radiology, 61,* 602 (1953).

5. C. L. Edwards and R. L. Hayes, *J. Nucl. Med., 10,* 103 (1969).

6. C. L. Edwards and R. L. Hayes, *J. Am. Med. Assoc., 212,* 1182 (1970).

7. W. W. Hunter, Jr., and H. W. de Kock, *J. Nucl. Med., 10,* 343 (1969).

8. J. A. Halperin, S. A. Stringer, and R. D. Eastep, in *Radiopharmaceuticals II,* Soc. Nucl. Med., New York, 1979, p. xxi.

9. M. W. Greene and W. D. Tucker, *Int. J. Appl. Radiat., 12,* 62 (1961).

10. C. Loc'h, B. Maziere, and D. Comar, *J. Nucl. Med., 21,* 171 (1980).

11. S. R. Vallabhajosula, J. F. Harwig, J. K. Siemsen, and W. Wolf, *J. Nucl. Med., 21,* 650 (1980).

12. S. W. Gunasekera, L. J. King, and P. J. Lavender, *Clin. Chim. Acta, 39,* 401 (1972).

13. R. L. Hayes, J. J. Rafter, B. L. Byrd, and J. E. Carlton, *J. Nucl. Med., 22,* 325 (1981).

14. S. M. Larson, J. S. Rasey, D. R. Allen, N. J. Nelson, Z. Grunbaum, G. D. Harp, and D. L. Williams, *J. Natl. Cancer Inst., 64,* 41 (1980).

15. M. J. Welch and T. J. Welch, in *Radiopharmaceuticals* (G. Subramanian, B. A. Rhodes, J. F. Cooper, and V. J. Sodd, eds.), Soc. Nucl. Med., New York, 1975, p. 73.

16. A. W. Ford-Hutchinson and D. J. Perkins, *Eur. J. Biochem., 21,* 55 (1971).

17. R. L. Hayes, B. L. Byrd, J. J. Rafter, and J. E. Carlton, *J. Nucl. Med., 21,* 361 (1980).

18. C. F. Baes, Jr. and R. E. Mesmer, *The Hydrolysis of Cations,* John Wiley and Sons, New York, 1976, Chap. 14.

19. R. L. Hayes, *Sem. Nucl. Med., 8,* 183 (1978).

20. S. E. Jones and S. E. Salmon, *Sem. Nucl. Med.*, *6*, 331 (1976).

21. P. Hoffer, *J. Nucl. Med.*, *21*, 394 (1980).

22. G. S. Johnston, *Int. J. Nucl. Med. Biol.*, *8*, 249 (1981).

23. F. H. DeLand, B. J. L. Sauerbrunn, C. Boyd, R. H. Wilkinson, Jr.,
     B. I. Friedman, M. Moinuddin, D. F. Preston, and R. M. Kniseley,
     J. Nucl. Med., *15*, 408 (1974).

24. H. D. Brereton, B. R. Line, H. N. Loder, J. F. O'Donnell, C. H.
     Kent, and R. E. Johnson, *J. Am. Med. Assoc.*, *240*, 666 (1978).

25. G. S. Johnston, M. F. Go, R. S. Benua, S. M. Larson, G. A.
     Andrews, and K. F. Hübner, *J. Nucl. Med.*, *18*, 692 (1977).

26. G. A. Andrews, K. F. Hübner, and R. H. Greenlaw, *J. Nucl. Med.*,
     19, 1013 (1978).

27. D. L. Longo, R. L. Schilsky, L. Blei, R. Cano, G. S. Johnston,
     and R. C. Young, *Am. J. Med.*, *68*, 695 (1980).

28. S. D. Richman, F. Applebaum, S. M. Levenson, G. S. Johnston,
     and J. L. Ziegler, *Radiology*, *117*, 639 (1975).

29. M. S. Milder, R. S. Frankel, G. B. Bulkley, A. S. Ketcham, and
     G. S. Johnston, *Cancer*, *32*, 1350 (1973).

30. S. D. Richman, J. N. Ingle, S. M. Levenson, J. P. Neifeld,
     D. C. Tormey, A. E. Jones, and G. S. Johnston, *J. Nucl. Med.*,
     16, 996 (1975).

31. R. S. Frankel, A. E. Jones, J. A. Cohen, K. W. Johnson, G. S.
     Johnston, and T. C. Pomeroy, *Radiology*, *110*, 597 (1974).

32. B. J. L. Sauerbrunn, G. A. Andrews, and K. F. Hübner, *J. Nucl.
     Med.*, *19*, 470 (1978).

33. R. L. Hayes and C. L. Edwards, in *Medical Radioisotope Scinti-
     graphy 1972*, Vol. II, IAEA, Vienna, 1973, p. 531.

34. F. Lomas, P. E. Dibos, and H. N. Wagner, *N. Engl. J. Med.*, *286*,
     1323 (1972).

35. R. D. Neumann, P. B. Hoffer, M. J. Merino, J. M. Kirkwood,
     A. Gottschalk, in *Medical Radionuclide Imaging 1980*, Vol. II,
     IAEA, Vienna, 1981, p. 475.

36. M. V. Merrick, S. W. Gunasekera, J. P. Lavender, A. D. Nunn,
     M. L. Thakur, and E. D. Williams, in *Medical Radioisotope
     Scintigraphy 1972*, Vol. II, IAEA, Vienna, 1973, p. 721.

37. R. H. Adamson, G. P. Canellos, and S. M. Sieber, *Cancer Chemo-
     ther. Rep.*, *59*, 599 (1975).

38. P. Hoffer, *J. Nucl. Med.*, *21*, 484 (1980).

39. R. L. Hayes, in *Tumour Localization with Radioactive Agents*,
     IAEA, Vienna, 1976, p. 29.

40. M. L. Thakur, in *Applications of Nuclear and Radiochemistry*
     (R. M. Lambrecht and N. Morcos, eds.), Pergamon Press, New
     York, 1982, p. 115.

41. E. V. Staab and W. H. McCartney, *Sem. Nucl. Med.*, *8*, 219 (1978).

42. C. Alberts, J. B. van der Schoot, and A. S. Groen, *Eur. J. Nucl. Med.*, *6*, 205 (1981).

43. E. H. Turbiner, S. D. J. Yeh, P. P. Rosen, M. S. Bains, and R. S. Benua,*Radiology, 127*, 437 (1978).

44. S. D. Richman, S. M. Levenson, P. A. Bunn, G. S. Flinn, G. S. Johnston, and V. T. DeVita, *Cancer, 36*, 1966 (1975).

45. T. L. Stahly and J. D. Shoop, *J. Nucl. Med., 16*, 1031 (1975).

46. R. L. Burleson, M. C. Johnson, and H. Head, *Ann. Surg., 178*, 446 (1973).

47. D. J. Hnatowich, W. W. Layne and R. L. Childs, *Int. J. Appl. Radiat. Isot., 33*, 327 (1982).

48. W. W. Layne, D. J. Hnatowich, P. W. Doherty, R. L. Childs, D. Lanteigne, and J. Ansell, *J. Nucl. Med., 23*, 627 (1982).

49. D. J. Hnatowich,*J. Nucl. Med., 17*, 57 (1975).

50. P. L. Hagan, G. E. Krejcarek, A. Taylor, and N. Alazraki, *J. Nucl. Med., 19*, 1055 (1978).

51. R. L. Hayes, J. E. Carlton, and Y. Kuniyasu, *Eur. J. Nucl. Med., 6*, 531 (1981).

52. D. J. Hnatowich and P. Schiegel, *J. Nucl. Med., 22*, 623 (1981).

53. J. Schuhmacher, W. Maier-Borst, and H. N. Wellman, *J. Nucl. Med., 21*, 983 (1980).

54. D. A. Goodwin, D. S. Stern, and H. N. Wagner, *J. Am. Med. Assoc., 206*, 329 (1968).

55. R. L. Hayes, in *The Chemistry of Radiopharmaceuticals* (N. D. Heindel, H. D. Burns, T. Honda, and L. W. Brady, eds.), Masson, New York, 1978, p. 155.

56. A. M. Dymov and A. P. Savostin, *Analytical Chemistry of Gallium,* Ann Arbor Science Publishers, Ann Arbor, 1970, p. 12.

57. K. Saito and M. Tsuchimoto, *J. Inorg. Nucl. Chem., 23*, 71 (1961).

58. M. H. Adatepe, M. Welch, E. Archer, R. Studer, and E. J. Potchen, *J. Nucl. Med., 9*, 426 (1968).

59. R. L. Hayes and D. H. Brown, in *Nuklearmedizin* (H. W. Pabst, G. Hor, and H. A. E. Schmidt, eds.), F. K. Schattauer Verlag, New York, 1975, p. 837.

60. B. Nelson, R. L. Hayes, C. L. Edwards, R. M. Kniseley, and G. A. Andrews, *J. Nucl. Med., 13*, 92 (1972).

61. R. L. Hayes, J. J. Szymendera, and B. L. Byrd, *J. Nucl. Med., 20*, 938 (1979).

62. S. M. Larson and G. L. Schall, *J. Am. Med. Assoc., 218,* 257 (1971).

63. J. Fogh, *Proc. Soc. Exp. Biol. Med., 138,* 1086 (1971).

64. J. F. Woessner, Jr., in *Lysosomes in Biology and Pathology,* Vol. 1 (J. T. Dingle and H. B. Fell, eds.), Chap. 11, Elsevier/ North-Holland, Amsterdam, 1969, p. 299.

65. R. L. Hayes and B. L. Byrd, in *Third International Radiopharmaceutical Dosimetry Symposium* (E. E. Watson, A. T. Schlafke-Stelson, J. L. Coffey, and R. J. Cloutier, eds.), Bureau of Radiological Health, Rockville, Md., 1981, p. 447.

66. R. E. Hartman and R. L. Hayes, *J. Pharmacol. Exp. Ther., 168,* 193 (1969).

67. R. E. Hartman and R. L. Hayes, *Fed. Proc., 26,* 780 (1967).

68. R. L. Hayes, J. E. Carlton, and B. L. Byrd, *J. Nucl. Med., 6,* 605 (1965).

69. T. Higasi, C. Akiba, Y. Nakayama, K. Ito, T. Hisada, T. Miki, and K. Kawai, *Jap. J. Nucl. Med., 8,* 163 (1971).

70. R. Sephton and J. Martin, *Int. J. Nucl. Med. Biol., 8,* 341 (1981).

71. W. P. Bradley, P. O. Alderson, W. C. Eckelman, R. G. Hamilton, and J. F. Weiss, *J. Nucl. Med., 19,* 204 (1978).

72. P. B. Hoffer, A. Samuel, J. T. Bushberg, and M. Thakur, *Radiology, 131,* 775 (1979).

73. R. L. Hayes, B. L. Byrd, J. J. Rafter, and J. E. Carlton, *J. Nucl. Med., 21,* 361 (1980).

74. P. A. G. Hammersley, D. M. Taylor, and S. Cronshaw, *Eur. J. Nucl. Med., 5,* 411 (1980).

75. R. L. Hayes and C. L. Edwards, *South Med. J., 66,* 1339 (1973).

76. D. C. Swartzendruber, B. Nelson, and R. L. Hayes, *J. Natl. Cancer Inst., 46,* 941 (1971).

77. D. H. Brown, D. C. Swartzendruber, J. E. Carlton, B. L. Byrd, and R. L. Hayes, *Cancer Res., 33,* 2063 (1973).

78. D. H. Brown, B. L. Byrd, J. E. Carlton, D. C. Swartzendruber, and R. L. Hayes, *Cancer Res., 36,* 956 (1976).

79. P. Bichel and H. H. Hansen, *Br. J. Radiol., 45,* 182 (1972).

80. Y. Ito, S. Okuyama, K. Sato, K. Takahashi, T. Sato, and I. Kanno, *Radiology, 100,* 357 (1971).

81. DeSales Lawless, D. H. Brown, K. F. Hübner, S. P. Colyer, J. E. Carlton, and R. L. Hayes, *Cancer Res., 38,* 4440 (1978).

82. R. L. Hayes and J. E. Carlton, *Cancer Res., 33,* 3265 (1973).

83. R. L. Hayes and J. E. Carlton, *J. Nucl. Med., 23,* P37 (1982).

84. A. W. Harris and R. G. Sephton, *Cancer Res.*, *37*, 3634 (1977).

85. R. Sephton, *Int. J. Nucl. Med. Biol.*, *8*, 323 (1981).

86. R. L. Hayes, D. H. Brown, J. R. Palisano, and S. P. Colyer, *Clin. Nucl. Med.*, *6*, D2 (1981).

87. P. Hoffer, *J. Nucl. Med.*, *21*, 282 (1980).

88. S. Menon, H. N. Wagner, and M.-F. Tsan, *J. Nucl. Med.*, *19*, 44 (1978).

89. L. J. Anghileri and M. Heidbreder, *Oncology*, *34*, 74 (1977).

90. E. J. Potchen, A. J. Elliott, B. A. Siegel, R. Studer, and R. G. Evans, *J. Surg. Oncol.*, *3*, 593 (1971).

91. K.-Y. Tzen, Z. H. Oster, H. N. Wagner, Jr., and M.-F. Tsan, *J. Nucl. Med.*, *21*, 31 (1980).

92. P. M. Gullino, *Progr. Exp. Tumor Res.*, *8*, 1 (1966).

93. H. S. Winchell, P. D. Sanchez, C. K. Watanabe, L. Hollander, H. O. Anger, J. McRae, R. L. Hayes, and C. L. Edwards, *J. Nucl. Med.*, *11*, 459 (1970).

94. S. R. Vallabhajosula, J. F. Harwig, and W. Wolf, *Int. J. Nucl. Med. Biol.*, *8*, 363 (1981).

95. L. J. Anghileri, *J. Nucl. Med. All. Sci.*, *22*, 101 (1978).

96. R. L. Hayes, J. J. Rafter, J. E. Carlton, and B. L. Byrd, *J. Nucl. Med.*, *23*, 8 (1982).

Chapter 18

ASPECTS OF TECHNETIUM AS RELATED TO NUCLEAR MEDICINE

Hans G. Seiler
Institute of Inorganic Chemistry
University of Basel, Switzerland

1. THE ELEMENT AND ITS ISOTOPES

1.1. Short Outline of History and Natural Abundance

When Mendelejeff published his periodic law in 1869, there was a gap
below manganese corresponding to the atomic number 43. This missing
element he called *eka-manganese* (Em) [1]. Since this time numerous
investigations have been undertaken to find this predicted element.
In 1925 Noddack et al. [2,3] believed they discovered it in concentrates of sperrylith, gadolinite, and columbite by means of x-ray
spectroscopic analysis. From the name of their homeland, they called
the new element "masurium".

Further research by different investigators showed these results
to be erroneous. In 1937 only the element 43 had actually been discovered by Perrier and Segrè in Italy [4]. They found it in a sample
of molybdenum, bombarded with deuterons in the Berkeley cyclotron,
which Lawrence sent to these investigators. As a result of this bombardment of Mo with deuterons, predominately the nuclear reactions
$^{94}_{42}Mo(d,n)^{95m}_{43}Tc$ and $^{96}_{42}Mo(d,n)^{97m}_{43}Tc$ occurred, rendering technetium the
first element to be produced artificially. This fact is the basis
of the name *technetium* (Greek: *technetos* = artificial) [5].

Since its discovery, searches for the element in terrestrial
materials have been made without success. From the 21 isotopes of
Tc known today none is stable, their half-lives being between a few
seconds and millions of years. If there had been a primordial occurrence, even the longest living isotope, ^{97}Tc, would have decayed.

Without any doubt the lithosphere contains ^{99}Tc produced by spontaneous fission of ^{238}U or by induced fission of ^{235}U with vagabonding
neutrons. As mineral enrichment will not take place, its concentration must be very small. On the other hand, Tc has been found in the
spectrum of S-, M-, and N-type stars. Its presence in stellar matter
is leading to new theories of the production of heavy elements in the
stars [6,7].

Until 1960 Tc was available only in small amounts and the price
was as high as \$2800/g. With the development of nuclear power plants
and the recycling of the burnt nuclear fuel elements, Tc is today produced in greater quantities and is offered commercially to holders of
Atomic Energy Commission permits at a price of about \$300/g. Today,
45 years after the discovery of Tc, its isotope ^{99m}Tc has become one
of the most frequently used radionuclides in nuclear medicine.

1.2. Radiophysical Properties of Different Isotopes

It remains to be determined if there is a stable isotope of Tc.
Answers to this question are given by the Mattauch rules [8] and
the empirical charge/mass relation. Mattauch rule IV states that
elements with an odd number of protons will have only one or two
stable isotopes. From the empirical charge/mass relation it can be
deduced that the stable mass numbers for element 43 would have to be
between 97 and 101. According to Mattauch rule II, stable nuclei of
even mass have even values of nuclear charge and an even number of
neutrons. By this, the only possible masses would be 97, 99, and
101. However, Mattauch rule I states that no stable isobar pairs
can exist with atomic numbers that differ by one unit. ^{97}Mo, ^{99}Ru,
and ^{101}Ru being stable, no stable isotope of Tc can exist.

Today, 21 isotopes and 7 nuclear isomers of Tc with mass numbers from 90 to 110 are known. Those with mass numbers from 90 to
97 decay by electron capture (EC) or by positron (β^+) emission,
whereas isotopes with masses from 98 to 110 are decaying by β^- emission (Table 1). In this way nuclides with mass numbers 92, 94, 95,

TABLE 1

Isotopes of Technetium

Isotope	Mode of decay	Half-life	γ Energies (keV)	Mode of production
^{90}Tc	β^+	50 sec	948; 1054; 511	^{92}Mo(p,3n)^{90}Tc
^{91}Tc	β^+, EC	3.3 min	218-2888; 511	^{92}Mo(p,2n)^{91}Tc
^{92}Tc	β^+, EC	4.4 min	329; 773; 511	^{92}Mo(d,2n)^{92}Tc
^{93m}Tc	IT, EC	43 min	392; 2645	^{92}Mo(d,n)^{93m}Tc; ^{92}Mo(p,γ)^{93m}Tc
^{93}Tc	β^+, EC	2.7 hr	1363; 1477; 511	^{92}Mo(d,n)^{93}Tc; ^{92}Mo(p,γ)^{93}Tc
^{94m}Tc	IT, β^+, EC	52 min	871; 1869; 511	^{93}Nb(α,3n)^{94m}Tc; ^{94}Mo(d,2n)^{94m}Tc
^{94}Tc	β^+, EC	4.88 hr	704; 650; 511	^{94}Mo(d,2n)^{94}Tc
^{95m}Tc	IT, β^+, EC	61 days	204; 786; 511	^{95}Mo(p,n)^{95m}Tc; ^{94}Mo(d,n)^{95m}Tc
^{95}Tc	EC	20 hr	766; 1074	^{95}Mo(p,n)^{95}Tc; ^{94}Mo(d,n)^{95}Tc
^{96m}Tc	IT, β^+, EC	52 min	34; 481; 778	^{93}Nb(α,n)^{96m}Tc; ^{96}Mo(d,2n)^{96m}Tc
^{96}Tc	EC	4.3 days	778; 850; 1127	^{96}Mo(p,n)^{96}Tc; ^{95}Mo(d,n)^{96}Tc
^{97m}Tc	IT	90 days	97	^{96}Ru(n,γ)^{97}Ru(EC)^{97m}Tc
^{97}Tc	EC	2.6×10^6 years		^{96}Ru(n,γ)^{97}Ru(EC)^{97}Tc

Isotope	Decay	Half-life	Gamma energies	Production
^{98}Tc	β^-	4.2×10^6 years	652; 745	^{98}Mo(p,n)^{98}Tc
^{99m}Tc	IT	6.02 hr	141	^{98}Mo(n,γ)^{99}Mo(β^-)^{99m}Tc
^{99}Tc	β^-	2.1×10^5 years		fission
^{100}Tc	β^-	15.8 sec	540; 591	^{99}Tc(n,γ)^{100}Tc
^{101}Tc	β^-	14.3 min	307; 532	^{100}Mo(n,γ)^{101}Mo(β^-)^{101}Tc
^{102m}Tc	IT, β^-	4.4 min	475; 628; 1103	^{102}Ru(n,p)^{102m}Tc; fission
^{102}Tc	β^-	5.3 sec	475; 628	^{102}Ru(n,p)^{102}Tc; fission
^{103}Tc	β^-	50 sec	136; 531; 884	Fission
^{104}Tc	β^-	18.1 min	360; 531; 884	Fission
^{105}Tc	β^-	7.6 min	108; 143; 322	Fission
^{106}Tc	β^-	36 sec	270; 2239	Fission
^{107}Tc	β^-	21.2 sec	103	Fission
^{108}Tc	β^-	5.17 sec		Fission
^{109}Tc	β^-	1.4 sec		Fission
^{110}Tc	β^-	0.82 sec		Fission

Source: Compiled from data from Ref. 9.

96, and 97 will yield on radioactive decay the corresponding stable
isotope of Mo, whereas those with mass numbers 98, 99, 100, 101, 102,
and 104 will transform to the corresponding stable isotope of Ru.

$$_{43}^{a}Tc \xrightarrow{\text{EC}/\beta^+} {}_{42}^{a}Mo \qquad \text{where } a = 92, 94, 95, 96, 97$$

$$_{43}^{b}Tc \xrightarrow{\beta^-} {}_{44}^{b}Ru \qquad \text{where } b = 98, 99, 100, 101, 102, 104$$

If there are nuclear isomers, the decay to the stable daughter will
proceed in two steps, each with its own half-life. In the cases
cited above, no other radioactive nuclides are involved and none
should be regarded in chemical or medical investigations.

The isotopes of Tc with mass numbers 93, 103, 105, 106, and
107 decay via more or less long chains of radioactive isotopes of
different elements, e.g.,

$$^{93m}Tc \xrightarrow[\text{43 min}]{\text{IT}} {}^{93}Tc \xrightarrow[\text{2.7 hr}]{\text{EC}/\beta^+} {}^{93m}Mo \xrightarrow[\text{6.9 hr}]{\text{IT}} {}^{93}Mo \xrightarrow[\text{3} \times 10^3 \text{ years}]{\text{EC}} {}^{93}Nb$$

$$^{103}Tc \xrightarrow[\text{50 sec}]{\beta^-} {}^{103}Ru \xrightarrow[\text{39.4 days}]{\beta^-} {}^{103m}Rh \xrightarrow[\text{56.1 min}]{\text{IT}} {}^{103}Rh$$

β^- decays are often accompanied by γ emissions, whereas β^+ radiation
is always followed by annihilation radiation, a γ emission of 511 keV.
Internal transition (IT) manifests by x-ray emission and conversion
electrons whereas electron capture decay (EC) is accompanied by the
"innere Bremsstrahlung" and x-ray emission.

1.3. Production and Availability of Different Isotopes

As can be seen from Table 1 (last row), isotopes of Tc may be produced by different nuclear reactions. Generally one can distinguish
three kinds of reactions:

1. Bombardment of suitable target nuclides with neutrons,
   protons, deuterons, or α particles
2. Neutron-induced fission of heavy nuclides such as Th, U, Pu
3. Production of a radioactive mother nuclide that on decay
   yields the isobar isotope of Tc.

Thus ^{102}Tc, ^{103}Tc, and ^{104}Tc can be obtained by bombarding the corresponding isotopes of Ru with fast neutrons, e.g.,

$$^{102}Ru + n \longrightarrow {}^{102}Tc + p \qquad [^{102}Ru(n,p)^{102}Tc]$$

For the bombardment with charged particles, the last must be of high kinetic energy. Therefore, these reactions can only be performed in accelerators. But even in great plants of this type, the particle flux and duration of bombardment being limited (compared with irradiation conditions with neutrons in nuclear reactors), only small quantities of the respective isotopes can be produced. In spite of this, some isotopes produced in this way are of interest as tracers in radiochemistry, e.g., ^{95m}Tc.

$$^{95}Mo + p \longrightarrow {}^{95m}Tc + n \qquad [^{95}Mo(p,n)^{95m}Tc]$$
$$^{94}Mo + d \longrightarrow {}^{95m}Tc + n \qquad [^{94}Mo(d,n)^{95m}Tc]$$

According to their kinetic energies, the same impinging particles may give rise to different reactions, e.g., $^{92}Mo(d,n)^{93}Tc$ and $^{94}Mo(d,2n)^{92}Tc$. As will be seen later, isotopes produced in these ways have no significance for studies in classical chemistry and are scarcely used in nuclear medicine.

The main source for the long-lived ^{99}Tc is the fission reaction of heavy nuclides such as U, Th, and Pu. For the fission of ^{235}U with thermal neutrons, ^{99}Tc has the highest yield (6.07%) among the fragments formed in this reaction. A 1000-MW reactor produces 28 g of ^{99}Tc daily. ^{99}Tc is extracted and purified from the waste solutions of the refining process of burnt nuclear fuels (Purex process).

Another important mode of production of Tc isotopes, above all of ^{99m}Tc, is the generation of a radioactive mother nuclide by one of the methods cited above, which on decay yields the corresponding isotope of Tc, e.g.,

$$^{98}Mo(n,\gamma)^{99}Mo \xrightarrow[66.02 \text{ hr}]{\beta^-} {}^{99m}Tc$$

$$^{235}U \text{ (fission) } {}^{99}Mo \xrightarrow[66.02 \text{ hr}]{\beta^-} {}^{99m}Tc$$

$$^{96}Ru(n,\gamma)^{97}Ru \xrightarrow[2.88 \text{ days}]{EC} {}^{97m}Tc$$

From the large group of Tc isotopes, only three are long-lived: ^{97}Tc, ^{98}Tc, and ^{99}Tc. But only the last is readily available because it is formed in the fission reaction. Therefore nearly all classical chemical work on Tc has been done with this isotope. One gram of ^{99}Tc corresponds to a radioactivity of 6.25 x 10^8 Bq = 0.017 Ci.

Of the short-lived isotopes, ^{99m}Tc is the most available in higher activity, since its mother nuclide, ^{99}Mo, can be easily produced by the n,γ reaction in nuclear reactors or isolated from fission products. Thus nearly all work with short-lived Tc isotopes has been done with ^{99m}Tc. This is the preferred isotope for use in nuclear medicine.

2. GENERATORS FOR THE PREPARATION OF ^{99m}Tc

The widespread application of ^{99m}Tc in nuclear medicine demands simple production methods of this isotope. These methods should allow production of the isotope in appropriate activities at reasonable costs and in a readily usable form of high purity. Furthermore, production should be close to the place of application because the half-life of ^{99m}Tc (6.02 hr) does not permit longer mailing times. These requirements can only be fulfilled by means of so-called generators.

The general concept of isotope generators is based on the separation of a short-lived daughter nuclide from the longer lived precurser. Usually the short-lived daughter is separated from the long-lived mother by chemical or physical means, whereby the mother nuclide is left behind. By disintegration of the mother nuclide fresh activity of the daughter is created and after a certain time the separation step may be repeated. The separation step is some-times called "milking" and correspondingly the mother nuclide remaining behind is referred to as the "radioactive cow" [10]. The first artificial radioisotope generator for medical diagnosis used the decay of ^{132}Te ($t_{1/2}$ = 3.2 days) to ^{132}I ($t_{1/2}$ = 2.3 hr) [11-14].

The ^{99m}Tc generator was first suggested for radiodiagnostic use in
1960 [15] and employed for the first time in 1961 [16].

The long-lived mother nuclide for ^{99m}Tc generators is ^{99}Mo
($t_{1/2}$ = 66.02 hr), which decays by β^- emission via a branched decay
to ^{99m}Tc (87.6%) and ^{99}Tc (12.4%) [9]. ^{99m}Tc ($t_{1/2}$ = 6.02 hr) trans-
forms by isomeric transition with the emission of a 140-keV γ radiation (11% internal conversion) to a basic level of ^{99}Tc. ^{99}Tc
($t_{1/2}$ = 2.14 x 10^5 years) disintegrates by β^- emission to the stable
^{99}Ru. As the half-life of ^{99}Tc is so much greater than that of ^{99m}Tc,
the activity of ^{99}Tc produced by the decay of ^{99m}Tc is negligible;
1 mCi ^{99m}Tc on decay yields 3.2 pCi ^{99}Tc.

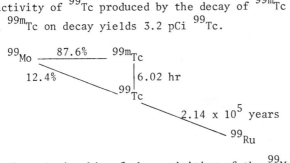

The relationship of the activities of the ^{99}Mo mother nuclide
and the ^{99m}Tc daughter are given by the decay law of a transient
equilibrium [17,18]. This can be described by the combination of
two differential equations for first order reactions. ^{99m}Tc is
formed according to the decay of ^{99}Mo and decays with its own half-
life. For the case of transient equilibrium, the decay constant of
the mother must be inferior to those of the daughter by a factor of
1/10 to 1/100. The growth of ^{99m}Tc in the system ^{99}Mo-^{99m}Tc is
described by the following equation [17]:

$$A_{99m_{Tc}} = 100 \, \frac{\lambda_B}{\lambda_B - \lambda_A} \left(1 - e^{-(\lambda_B - \lambda_A)t} \right) \delta_{A/B}$$

where $A_{99m_{Tc}}$ = activity of ^{99m}Tc in percent of activity of ^{99}Mo
at time t

λ_A = decay constant of ^{99}Mo

λ_B = decay constant of ^{99m}Tc

$\delta_{A/B}$ = branching ratio, characterizing the amount of ^{99}Mo
decaying to ^{99m}Tc and ^{99}Tc

t = time elapsed since last ^{99m}Tc-^{99}Mo separation

Substitution for the corresponding constants reduces the equation to

$$A_{99m_{Tc}} = 94.56*(1 - e^{-0.1045*t})$$

The time which must elapse between two separations to obtain maximum activity of ^{99m}Tc can be calculated from the following equation:

$$t_m = \frac{1}{\lambda_B - \lambda_A} \ln \frac{\lambda_B}{\lambda_A}$$

After substitution of the constants and calculation, it can be seen that maximum ingrowth of ^{99m}Tc has happened after 22.9 hr.

It must be mentioned that ^{99m}Tc produced in generators is never free from carrier, in this case ^{99}Tc, which is produced by the isomeric transition of ^{99m}Tc and the direct decay of ^{99}Mo. Due to the enormous difference in half-lives an accumulation of ^{99}Tc takes place in generators not milked for days, so that chemically significant quantities of this isotope can be formed. Thus 24 hr after the last separation the ratio $^{99}Tc/^{99m}Tc$ is 2.5, but after 72 hr 11.88 [19]. As will be seen later, this accumulated ^{99}Tc can be responsible for poor labeling results [20,21]. Therefore it is best to discard the first milking solution of a new generator or from one not milked for days [22].

The main requirements for a ^{99m}Tc generator system usable in nuclear medicine include [23]:

1. The device must be safe, convenient, simple to handle, and rapid to operate. It should yield ^{99m}Tc fractions of very high purity and in a small final product volume.

2. The solution should be sterile, free of pyrogens, and suitable for direct intravenous injections; it should require little further processing for tagging colloids and chelating agents.

3. Conditions with respect to shielding, packaging, scheduling, and shipping must be as efficient and economical as possible.

Currently there are three types of generators in use for the production of ^{99m}Tc, which differ above all in the separation technique: chromatographic, sublimation, and extraction systems. All

three systems can be operated with different starting material: ^{99}Mo
from fission of ^{235}U and ^{99}Mo produced by n,γ reaction with ^{98}Mo.
The evaluation of a system should be made with respect to the starting material, separation efficiency, attainable specific activity,
as well as radionuclidic and chemical purity. The main requirements
for generator systems cited above must be fulfilled.

2.1. The Starting Material: ^{99}Mo

The first step in the preparation of a ^{99m}Tc generator will be the
production of the starting material. The fission of ^{235}U yields
relatively large quantities of product with high specific activity
of ^{99}Mo, but special installations are needed for pile irradiation,
chemical processing, and waste disposal [24]. The main advantages
of the activation reaction are the high nuclidic purity of the
product and the absence of a postirradiation chemical treatment.
On the other hand, the attainable specific activity is low. For the
production of ^{99}Mo by the fission reaction, metallic U, U-Al alloys,
or UO_2 are irradiated with neutrons [25-27]. Most commonly enriched
U targets (^{235}U content up to 93%) are used in order to suppress the
production of ^{239}Pu by activation of ^{238}U. By the bombardment of
^{235}U with thermal neutrons, ^{236}U is formed and undergoes spontaneous
fission. In this fission reaction more than 200 different fission
nuclides are produced, more than half of them radioactive [28]. The
independent fission yield of ^{99}Mo in this reaction is very low. It
is mainly produced in the mass 99 fission product chain [29].

$$^{99}Zr \xrightarrow[33 \text{ sec}]{\beta^-} {}^{99}Nb \xrightarrow[2.4 \text{ min}]{\beta^-} {}^{99}Mo$$

The total fission yield of ^{99}Mo is 6.06%. In the fission reaction
three stable isotopes of Mo are also produced with different fission
yields: ^{97}Mo (6.2%), ^{98}Mo (5.9%), and ^{100}Mo (6.5%) [28]. The specific
activity of ^{99}Mo is determined by the irradiation time, neutron flux,
and decay period after reactor discharge. Today, specific activities
up to 30,000 Ci/g are feasible [19]. For safety of handling such

high activities so-called hot laboratories are indispensable. After
irradiation the target must be processed for recovery and purification of the produced [99]Mo. This processing can be accomplished by
wet chemical treatment or by sublimation of MoO$_3$.

In the wet treatment the irradiated target is dissolved in acid
media, whereby gaseous fission products, in particular ^{133}Xe, are
released. By oxidation Mo is quantitatively converted into molybdate.
Addition of NaOH precipitates U and insoluble fission product hydroxides. The filtrate is acidified with nitric acid and most of the
iodine is volatilized [26]. The acid solution is treated with alumina
in order to bind molybdate and other nuclides to the alumina. Washing
with neutral or weakly alkaline solutions eliminates most other salts
and remaining iodine. After this, molybdate is brought into solution
with NaOH whereas other nuclides, e.g., Te, remain in the alumina
phase. Final purification is accomplished by cation-exchange chromatography. In practice, this isolation and purification of [99]Mo is a
much more complex procedure because some of the steps must manifoldly
be repeated. There are also other wet purification processes which
use selective precipitation or extraction of Mo [30,31]. With regard
to safety and procedural simplicity, multiple-step separation methods
are complicated especially in the presence of volatile fission products,
the use of corrosive acids and bases, and the need to treat radioactive
liquid wastes.

A much simpler method to separate [99]Mo from irradiated UO$_2$ makes
use of the volatility of MoO$_3$ at higher temperatures [32]. The irradiated UO$_2$ target is heated to about 700°C in a O$_2$ stream. By this UO$_2$
is converted to U$_3$O$_8$ and Mo to MoO$_3$. On further heating to 1300°C
and evacuation, MoO$_3$ sublimes from the target and can be condensed
on the surface of a cooled Pt collector from which it is dissolved
with aqueous NH$_3$. The solution, still contaminated with [103]Ru and
[132]Te, must be purified by chromatography on an alumina column or by
selective precipitation with α-benzoinoxime. The final alkaline
solution of Mo showed a purity better than 99.99% in a yield exceeding 90% [33].

A very effective separation and purification of ^{99}Mo from a
fission product mixture by a single adsorption-desorption process
uses silver-coated carbon granules as adsorbent [34].

The main radiocontaminants in routinely produced ^{99}Mo by the
fission reaction are ^{131}I, ^{103}Ru, ^{106}Ru, ^{106}Rh, ^{95}Zr, ^{95}Nb, and ^{132}Te.
The activities of these radionuclides are in the ppm to sub-ppm range
with respect to ^{99}Mo. For the production of ^{99}Mo by neutron activation, analytical grade molybdenum trioxid, ammonium molybdate, or
molybdenum metal is irradiated in a reactor with high neutron flux.
The cross-section for the nuclear reaction ^{98}Mo$(n,\gamma)^{99}$Mo with thermal
neutrons is 0.51 barn. By the irradiation of complex compounds of
Mo such as molybdenum hexacarbonyl [35], molybdenum oxinate [36], or
molybdenum phthalocyanine [37], advantage can be taken of the Szillard-
Charlmers reaction in order to produce enriched ^{99}Mo.

Recently, eutectic mixtures of PbO and MoO$_3$ have been used as
targets for irradiation [38]. Irradiated targets of molybdenum trioxid or ammonium molybdate dissolved in aqueous NH$_3$ can be directly
used as feed solution for ^{99m}Tc generators without further purification. The attainable specific activities of ^{99}Mo are in the order
of 4 Ci/g (natural Mo with 23.8% ^{98}Mo) to 15 Ci/g (enriched Mo with
93% ^{98}Mo) [28]. Activation of other Mo isotopes presents no serious
problems because their half-lives are short. As it can be seen from
the attainable specific activities of ^{99}Mo, ^{99m}Tc generators fed with
^{99}Mo produced by neutron activation will be of limited total activity
or low eluate concentration. In spite of the high nuclidic purity
and the absence of a postirradiation treatment in the case of n,γ-
produced ^{99}Mo, today preference is given to fission-produced ^{99}Mo as
starting material for the preparation of ^{99m}Tc generators.

2.2. Types of ^{99m}Tc Generators

Because the mother nuclide ^{99}Mo and the daughter ^{99m}Tc are isotopes
of two different elements with quite different chemical properties,
all radiochemical separation methods can be used for the construction

of ^{99m}Tc generators, such as chromatography, extraction, and sublimation.

2.2.1. *Chromatographic Generators*

The principle of chromatographic generators is based on the fact that
molybdate ions are much more strongly bound to alumina than pertechnetate ions. As a result there is a marked difference between the
two ions with respect to partitioning between their aqueous solutions
and solid alumina. The principle of construction of the generator is
very simple. It consists of a glass column filled with acidic aluminium oxide and an appropriate elution device. ^{99}Mo in high specific
activity, produced by one of the methods described above, is adsorbed
on the alumina column in the form of molybdate ions. The ^{99m}TcO$_4^-$
ions formed in the decay of ^{99}Mo can be eluted from the column by
diluted mineral acids or isotonic NaCl solution [39]. The break-
through in the use of commercial ^{99m}Tc generators came with the first
closed or sterile generator, developed at the Squibb Institute for
Medical Research in 1966. Such a closed system consists of an alumina
column fed with ^{99m}MoO$_4^{2-}$ in high specific activity, a closed admit-
tance system for the elution solution, and a collecting device equipped
with a 0.22-μm membrane filter. Using evacuated collecting vials, the
eluent is sucked through the column to produce a sterile and apyrogenic
solution of ^{99m}TcO$_4^-$. Today, most of the commercial generators are
of this type. Other systems use pressurized tanks for the eluent.

Among the different adsorbents such as hydrous zirconium oxide,
manganese dioxide, ferric oxide, charcoal, silica gel, etc., acidic
alumina is still the preferred sorption medium. A column of 3.5 g
Al$_2$O$_3$ has a capacity to bind more than 21 mg Mo, a quantity that
corresponds to 2.1 x 10^3 Ci of ^{99}Mo. But generators with such high
activity could not be built. The specific activity of ^{99}Mo on the
alumina should not be too high because the high radiation dose per
unit volume, mainly produced by the β^- emission of ^{99}Mo, greatly
increases the incidence of undesired side reactions. Therefore the
tendency in the development of new generator columns is to achieve

a diffuse distribution of ^{99}Mo within the column [40]. The elution profiles of generators, plots of specific activity of the eluate as function of eluate volume, depend on the bed size of the alumina column. With increasing bed size the peak heights decrease whereas the peak widths increase. Thus, small column size is necessary for high specific activities in the eluate. The minimum column volume is directly connected with the mode of ^{99}Mo production. Only fission-produced ^{99}Mo (almost carrier-free) permits the use of small-size columns. This is one of the main reasons that preference is given to fission product ^{99}Mo in the preparation of commercial ^{99m}Tc generators [19].

On milking chromatographic generators it can often be observed that ^{99m}Tc is incompletely removed from the column. The degree of elution is expressed by the so-called elution yield, the percentage of ^{99m}Tc eluted from the total amount present in the generator. In most cases the elution yield is significantly less than 100%. This can be explained by a reductive environment created by the exposure of water to a high radiation dose. Therefore TcO_4^- is reduced to lower oxidation states of Tc. The reduced species are strongly bound to alumina and cannot be eluted with isotonic NaCl solution [41]. Generally, the elution yield of a generator decreases with increasing number of elutions [42]. In high-activity generators an activity-related decrease in elution yield was observed [43].

Thus, for the preparation of chromatographic generators the activity of ^{99}Mo should be below a threshold level determined by the size of the column and the kind of distribution of the ^{99}Mo activity within the column [40]. Elution yields exceeding 90% can be obtained from columns with a ^{99}Mo activity below 250 mCi. Another improvement that will result in higher elution yields is the addition of inorganic or organic oxidants to the elution solution, but this may interfere with the subsequent preparation of ^{99m}Tc-labeled radiopharmaceuticals. Another method to ensure the desired high specific activity of ^{99m}Tc in the eluate is to overload the generator with ^{99}Mo to such an extent that the loss in eluted ^{99m}Tc activity will not matter.

2.2.2. Solvent Extraction Generators

The principle of solvent extraction generators is based on the great
difference in distribution constants of compounds of Mo and Tc
between aqueous and organic phases. Widespread investigations on
the extraction behavior of Tc and Mo showed that alcohols, ketones,
and tri-n-butylphosphate are the best extractants for Tc from acid
solutions, whereas for basic solutions ketones and cyclic amines
were most effective [44,45]. Most frequently, methylethylketone and
methylisobutylketone served as extractants for ^{99m}Tc from basic solutions of MoO_4^{2-} [46]. ^{99m}Tc can also be extracted from alkaline solutions of irradiated Mo by tetraphenylarsonium chloride in chloroform
[47]. In these systems ^{99m}Tc is extracted in the organic phase,
mostly as pertechnetate. After phase separation, the organic phase
is purified by passing an alumina column and evaporated. The remaining ^{99m}Tc is of high specific activity and levels of chemical and
nuclidic impurities are lower than in eluates from chromatographic
generators [48]. The aqueous phase can be milked repeatedly after
new ingrowth of ^{99m}Tc. Several liquid extraction generators have
been constructed [49,50]. This type of generator is especially useful if ^{99}Mo produced by neutron activation serves as starting material. Even with a starting material of low specific activity, high
specific activities of ^{99m}Tc can be reached.

2.2.3. Sublimation Generators

One of the most effective principles for the separation of Tc from
Mo is sublimation. By 1937 Perrier and Segrè had used the different
volatilities of Tc_2O_7 and MoO_3 to separate Tc from Mo [51]. Tc_2O_7
volatilizes completely in an O_2 stream at temperatures above 310°C,
whereas MoO_3 needs heating above 1280°C for sublimation. Generally,
a sublimation generator consists of a tube packed with $^{99}MoO_3$ or a
mixture of $^{99}MoO_3$/silica which is placed in a furnace. A carrier gas
(O_2 or air) passes through the heated tube. At the cold parts of
the apparatus, the volatilized Tc_2O_7 condenses and can be rinsed in
a collection vial with isotonic NaCl solution to give a colorless

solution of Na^{99m}TcO$_4$. The yield using MoO$_3$ of low specific activity is 25-30% [52]. The yield increases markedly up to 80% when fission-produced MoO$_3$ with high specific activity is used [53]. A quite different kind of sublimation generator uses a molten eutectic mixture of PbO^{99}MoO$_3$. By bubbling O$_2$ through the smelt, Tc$_2$O$_7$ is carried off with the gas and can be collected. Yields up to 67% can be obtained even with a n,γ-produced ^{99}Mo [54]. The purity of sublimed ^{99m}Tc$_2$O$_7$ is very high, especially with n,γ-produced ^{99}Mo as starting material.

2.3. Quality of the Separated ^{99m}Tc

Radioactive materials used for in vivo diagnosis or therapy are generally considered to be drugs [55]. In addition to the general characteristics of a pharmaceutical product such as chemical purity, apyrogenicity, and sterility, radiopharmaceuticals must be considered with respect to their radionuclidic and radiochemical purity.

Radionuclidic purity relates to the presence of radioisotopes other than the stated radionuclide. The kind and level of radionuclidic impurities in the milkings of ^{99m}Tc depend on the mode of preparation of ^{99}Mo as well as the separation technique used for the generator. Standards for the qualities of radiopharmaceuticals are given by the national pharmacopoeias. Thus, USP radiopharmaceuticals criteria specify that the maximum levels of radionuclidic impurities allowed at the time of administration to the patient per mCi of ^{99m}Tc are 1 μCi ^{99}Mo, 0.05 μCi ^{131}I, 0.05 μCi ^{103}Ru, and 0.1 μCi for the sum of all other γ-emitting isotopes. The maximum ratio of gross α emitters to ^{99m}Tc must be less than 10^{-9}. Special attention must be given to generators used for a longer time. Even if the eluates from the fresh generator meet the specifications given by the pharmacopoeias, with the passing of time the ratio of the long-lived impurities to ^{99m}Tc may exceed the legal specifications.

Radiochemical purity reflects the percentage of the radionuclide present in the stated chemical form. Eluates of ^{99m}TcO$_4^-$

are preferably tested for radiochemical purity by paper or thin-layer chromatography [56,57]. By these methods TcO_4^- can be easily separated from lower oxidation states of Tc. The chemical purity of ^{99m}Tc separations relates to the presence of substances other than those specified by the mode of preparation. An impurity often encountered in the eluates from chromatographic generators is aluminum, eluted in soluble and insoluble form from the alumina column. It has been found in µg/ml levels [58]. Insoluble aluminum is often accompanied by ^{99}Mo. Both can be filtered. Therefore chromatographic generators must be equipped with an effective membrane filter in the outlet. In the milkings of extraction or sublimation generators, problems with chemical impurities are scarcely encountered.

Finally, it must be said that because of the ease of handling, the commercial availability, the obtainable high-volume activity, and the sterility and apyrogenicity of the eluate, the chromatographic generator utilizing ^{99}Mo from uranium fission is the most widely used type of ^{99m}Tc generator. Further, radioactive hazards to personnel are minimized by these closed systems in comparison with the other types of generators for which solutions of high ^{99}Mo activity must be handled.

3. CHEMISTRY OF TECHNETIUM

When discussing the chemistry of technetium, it must be remembered that there exists no stable isotope of this element. Therefore all classical chemical work, preparative and analytical, has been done with the most readily available long-lived ^{99}Tc. To minimize radioactive hazards and to overcome the problems related to wastes, only small quantities of this isotope can be used; 1 mCi ^{99}Tc = 59.2 mg = 5.98×10^{-4} mol.

The chemistry of technetium is very similar to that of rhenium, but there is a remarkable difference in the behavior of manganese, although stoichiometric similarities in TcO_4^- and MnO_4^- as well as in the metal carbonyls can be observed. Tc adopts oxidation states from -I to VII in its compounds, the IV and VII states being the most important ones. The III and V oxidation states can be fairly stabilized by a number of chelating agents. Although Re and, to a smaller extent, Tc show a tendency to form Tc-Tc bonds in the lower oxidation states, the formation of such compounds has never been observed for Mn. The chemistry of solvated cations of Tc, comparable to Mn^{2+}, is unknown. The ground state of Tc has an outer electron configuration $4s^2 4p^6 4d^6 5s^1$ ($S = 9/2$) [59].

Elemental Tc is a silver-grey metal crystallizing in the hcp arrangement, melting at 2140°C, and has a density of 11.5. The metal burns in oxygen above 400°C to give the subliming Tc_2O_7. In moist air the metal is slowly oxidized to $HTcO_4$. Dissolution of the metal in concentrated HNO_3 or hot concentrated H_2SO_4 also yields $HTcO_4$. The metal is not dissolved by HF or HCl. In the following, some examples of Tc compounds of the different oxidation states will be described.

3.1. Tc(-I)

Species of Tc(-I) are obtained by reduction of NH_4TcO_4 with potassium in ethylenediamine [60]. The -I oxidation state is stabilized by carbonyl ligands as in $Na[Tc(CO)_5]$, prepared by the reduction of $Tc_2(CO)_{10}$ in tetrahydrofuran with sodium amalgam [61].

3.2. Tc(0)

Representatives are the carbonyls $Tc_2(CO)_{10}$ [62] and $TcRe(CO)_{10}$ [63], which have octahedral geometry.

3.3. Tc(I)

The low oxidation state Tc(I) can be stabilized by strong π-back bonding ligands such as carbon monoxide, cyanide, aromatic hydrocarbons, etc. Examples are $Tc(CO)_5X$ (X = Cl, Br, I) [61]; the large group of compounds of the type $Tc(CO)_3L_2Cl$ [L = PPh_3, $AsPh_3$, $SbPh_3$, pyridine, 1/2(1,10-phenanthroline), isonitrile], as well as the dithio-carbamates $Tc(CO)_4(S_2CNR_2)$ (R = Me, Et) [64]. Also several cations such as $[Tc(CO)_4(PPh_3)_2]^+$ [64], cis- and trans-$[Tc(CO)_2\{P(OEt)_2Ph\}_4]^+$ [65], and $[Tc(C_6H_6)_2]^+$ [66,67] have been described. An anionic compound $[Tc(CN)_6]^-$ is obtained from the reduction of TcO_4^- by potassium amalgam in the presence of KCN [68]. Most of these complexes have octahedral geometry, sometimes a distorted one.

3.4. Tc(II)

$Tc(diars)_2X_2$ (X = Cl, Br, I) is the first compound isolated of Tc(II) [69]. Complexes of the type $TcX_2[PPh(OEt)_2]_4$ react with carbon monoxide at atmospheric pressure to give cis- and trans-dicarbonyl isomers by replacement of the halides [65]. $(nBu_4N)[Tc(NO)Br_4]$, which can be prepared in high yield [70], can serve as starting material to produce compounds such as $(nBu_4N)[Tc(NO)Cl_4]$ and $(nBu_4N)_2[Tc(NO)(CNS)_5]$. An organometallic compound, the cyclopentadienyl Tc(II) dimer $[Tc(C_5H_5)]_2$, has also been reported [71].

3.5. Tc(III)

In the III oxidation state Tc complexes with a variety of ligands in different coordination geometries. The ligands should be good π acceptors such as PPh_3, diars, or carbonyl. The first isolated compounds of Tc(III) were mixed complexes with diars and halides as ligands, $[Tc(diars)_2X_2]X$ (X = Cl, Br, I) [69]. An analogous compound is formed with diphos, $[Tc(diphos)_2Cl_2]Cl$ [72]. These complexes show a trans-octahedral arrangement of the ligands. A compound mer-$Tc(Me_2PhP)_3Cl_3$ is formed by the reaction of $(NH_4)_2[Tc(Cl)_6]$ with Me_2PhP in absolute ethanol [73]. Tc(III) species in solution have been produced by the reduction of TcO_4^- with $SnCl_2$ in the

presence of polydentate organic ligands such as citrate [74] or DTPA
[75]. The existence of Tc-Tc bonds have been proved for the dimeric
complexes $[Tc_2(O_2CCMe_3)_4Cl_2]$ and $(nBu_4N)_2[Tc_2Cl_8]$ [70,76], but also
in species containing Tc in fractional oxidation states between II
and III, e.g., $[Tc_2Cl_8]^{3-}$ [77] and $[Tc_2(OC_5H_4N)_4Cl]$ [78].

3.6. Tc(IV)

IV is one of the most important oxidation states of Tc. Tc(IV) forms
binary compounds such as the oxide, the sulfide, and the halides.
The black TcO_2, obtained by thermal decomposition of NH_4TcO_4, has a
distorted rutile structure isotypic to that of MoO_2. The sparingly
soluble hydrated dioxide $[TcO_2 \cdot 2H_2O]_n$, sometimes used as starting
material for the preparation of Tc(IV) complexes, can be obtained by
the hydrolysis of $[TcX_6]^{2-}$ or the reduction of TcO_4^- with Zn in dilute
mineral acid [79]. TcS_2, produced by heating Tc_2S_7 with sulfur in a
vacuum, is mostly nonstoichiometric. The red-brown, paramagnetic
$TcCl_4$ has a structure similar to that of $ZrCl_4$, i.e., there are
linked $TcCl_6$ octahedra [79,80]. The hexahalogenotechnetates(IV)
$[TcX_6]^{2-}$ (X = Cl, Br, I) [80] are commonly employed as starting
materials for the preparation of other Tc(IV) compounds. The chloro
and bromo complexes are readily synthesized by the reduction of TcO_4^-
in hot solutions of the corresponding hydrohalic acid, whereas the
iodo complex is obtained from the chloro or bromo species by evaporation in HI [81]. The relative stabilities of these complexes are
Cl > Br > I [81]. Crystal structure of $(NH_4)_2[TcCl_6]$ shows the complex to be octahedral [82]. $K_2[Tc(CN)_6]$ [83] and ternary halide
complexes such as $[Tc(Ph_3P)_2Cl_4]$, $[Tc(Ph_3As)_2Cl_4]$ [84], and
$[Tc(bipy)Cl_4]$ [72] have been published.

Tc(IV) complexes with such diphosphonate ligands as HEDP are
of special interest for the preparation of radiodiagnosticals for
imaging bony tissue [87-89]. In acidic aqueous perchlorate solutions Tc(IV) shows a charge of 2+, which proves the existence of
TcO^{2+} or $Tc(OH)_2^{2+}$ [85,86].

3.7. Tc(V)

Several halides and oxohalides of Tc(V), such as TcF_5 and $TcOCl_3$
[79,80], as well as hexahalogenotechnetates(V) are known, e.g., the
alkali metal salts of $[TcF_6]^-$ [90] and $[Tc(CNS)_6]^-$ [80]. Recently,
a large number of Tc(V) complexes have been prepared and characterized by x-ray diffraction. It could be shown that the previously
synthesized $[TcOCl_4]^-$ ion [91,92] has a square pyramidal geometry,
the Tc atom being displaced from the plane of the chlorines toward
the oxygen [93]. The analogous complexes $[TcOBr_4]^-$ [94] and $[TcOI_4]^-$
[95] have also been produced. The unit TcO^{3+} exists in a variety
of complexes with organic ligands. Thus $[TcO(SCH_2CH_2S)_2]$ [96],
$[TcO(SCH_2COS)_2]^-$ [97], and $[TcO(SCH_2CH_2O)_2]^-$ [98], all prepared
from $[TcOCl_4]^-$ by ligand exchange, show square pyramidal structures.
Chelating agents are able to form stable complexes with Tc(V) in
aqueous solution, whereas ions with monodentate ligands must usually
be prepared in nonaqueous media. With most common nitrogen ligands
Tc(V) tends to adopt an octahedral geometry by incorporating trans-
dioxotechnetium(V), $tr\text{-}TcO_2^+$. This structure has been confirmed by
x-ray diffraction for $tr\text{-}TcO_2(cyclam)ClO_4 \cdot H_2O$ [99]. Other complexes
containing $tr\text{-}TcO_2^+$ have been synthesized with ethylenediamine [100],
diaminopropane [101], cyanide [102], and pyridine [103].

Complexes of Tc(V) with oxygen-containing ligands, such as
citrate, manitol, and glucoheptonate, were evaluated early as renal
imaging agents [104]. As could be shown by the structure determination of $(nBu_4N)[TcO(cat)_2]$ [105], Tc(V) is coordinated in a square
pyramidal geometry. It can be assumed that this geometry is also
used for complexes with ethylene glycolate and carbon hydrate ligands
[106]. In the reduction of TcO_4^- with Ti(III) [90] and Sn(II) in the
presence of citrate [75], as well as in TcO_4^- solutions in cold concentrated HCl [107], Tc(V) species are claimed to be transient inter-
mediates.

3.8. Tc(VI)

Only a few compounds of Tc(VI) are known. The halide TcF_6 [108] and
the oxohalides $TcOF_4$ [109] and $TcOCl_4$ [110] have been reported. The
isolation of $TcCl_6$ has been disputed [110]. Cathodic reduction of
TcO_4^- in acetonitrile solution containing tetramethylammonium salts
yields violet crystals of $(Me_4N)_2TcO_4$ [111]. There is evidence for
the intermediate formation of TcO_4^{2-} from electrochemical reduction
of TcO_4^- in alkaline aqueous media [112].

3.9. Tc(VII)

Although there are only a few compounds known, VII is the most important and stable oxidation state of Tc. The heptoxide Tc_2O_7 is a yellow crystalline solid (mp 119.5°C) consisting of molecules in which
TcO_4 tetrahedra share an oxygen atom and the Tc-O-Tc chain is linear
[79]. The volatility of the heptoxide is utilized in sublimation
generators for [99m]Tc. The black heptasulfide Tc_2S_7 is obtained by
saturation of hydrochloric acid solutions of TcO_4^- with H_2S. The
precipitation is often incomplete [79]. Simple halides of Tc(VII)
are unknown, but oxohalides such as TcO_3F [113] and TcO_3Cl [110]
have been published. The most important compounds are the pertech-
netates of monovalent alkali metals, $MTcO_4$, which are the starting
materials for nearly all preparations of technetium compounds.

4. ANALYTICAL CHEMISTRY OF TECHNETIUM

The widespread use of ^{99m}Tc in nuclear medicine and the related in
vitro investigations with ^{99}Tc demanded the development of effective
analytical methods. These methods should determine the element in
the micro- and picogram region in a variety of compounds. The applicability of such classical chemical methods as volumetry and gravim-
etry are limited to those problems whereby rather large quantities

of Tc are to be determined, as in the processing of spent nuclear
fuels. Generally, the methods applicable to the analytical problems
in nuclear medicine and chemical investigations are radiometric
methods; mass spectrometry; emission spectrometry; IR, UV, and visible spectrophotometry; voltammetric methods; and nuclear magnetic
resonance. Sometimes efficient separation steps must preceed the
final determination step. According to the statements made in Sec.
1.3, the following considerations are limited to ^{99}Tc and ^{99m}Tc.

4.1. Radiometric Methods

^{99}Tc can be characterized and quantitatively determined by measuring
its low-energy β^- emission (E_{max} = 292 keV). The specific activity
of ^{99}Tc is 630 disintegrations sec^{-1} μg^{-1}. Thus, determinations in
the nanogram range can be accomplished using thin end-window Geiger-
Müller counters [114], windowless flow counters, or liquid scintillation counters [114-116], the latters being the most sensitive (counting efficiency up to 94%). In the presence of other radioactive
nuclides in the sample to be analyzed, ^{99}Tc can be separated by
selective solvent extraction or chromatography.

    ^{99m}Tc can be measured with much greater sensitivity, the spe-
cific activity being 1.95 x 10^{11} disintegrations sec^{-1} μg^{-1}. The
soft, partially converted γ emission may be measured by Geiger-Müller
counters, x-ray proportional counters, or, preferably, by γ-scintilla-
tion detectors [NaI(Tl) well type] in combination with a γ-spectrom-
eter [117]. In the presence of other γ-emitting radionuclides, semiconductor detectors (GeLi) are preferred because of their better energy
resolution. Using well-type NaI(Tl) detectors (counting efficiency
about 40%), less than 10^{-17} g of ^{99m}Tc can be determined. For the
quality control of ^{99m}TcO$_4^-$ eluates and ^{99m}Tc-labeled pharmaceuticals,
chromatographic scanners equipped with scintillation detectors or
proportional counters are used to localize ^{99m}Tc on paper or thin-
layer chromatograms. Radiometric measurement is the only method
applicable to ^{99m}Tc. Quantities in the microgram and nanogram range

as used in chemical analytical methods would present serious radiation hazards.

Neutron activation analysis was also used for the determination of ^{99}Tc by the reaction ^{99}Tc$(n,\gamma)^{100}$Tc. Since the half-life of ^{100}Tc is very short (15.8 sec), special installations for reactor discharge and measurement are necessary. Pre- and postirradiation separation and purification steps of Tc (extraction, distillation, ion exchange) are indispensable [118]. The detection limit is in the nanogram range.

4.2. Mass Spectrometric Methods

Mass spectrometric methods for ^{99}Tc stand for high specifity and sensitivity, the detection limit being in the nanogram range [119]. However, this method requires considerable efforts for preparation and purification of the samples. Potential interferences by isobares, e.g., ^{99}Ru, the daughter of ^{99}Tc, must be excluded. The purified solution of Tc, mostly as TcO_4^-, is deposited on an iridium emitter. After evaporation, Tc is reduced to the metal in a stream of hydrogen. Tc^+ ions are produced by electron bombardment or thermal ionization. Special methods have been developed for the determination of ^{99}Tc in the subpicogram range in different matrices [120].

4.3. Optical Emission Spectrometry

Optical emission spectrometry, arc and spark, is a very efficient analytical method of good sensitivity (down to 0.1 µg Tc) and specificity [121,122]. The principal lines for arc and spark spectra of neutral Tc atoms and Tc^+ are between 254.324 and 429.706 nm [123]. The line with the least interference in the arc spectra is 403.163 nm; the most intense line of the spark spectra is 254.324 nm. Atomic absorption spectral methods with electrothermal graphite furnace for the determination of Tc have been published, but hollow cathode lamps of Tc are not yet commercially available [124,125].

4.4. Infrared Spectrometry

An infrared spectrometric method for the determination of Tc has
been published by Magee and Al-Kayssi [126]. It is based on the
strong absorption of tetraphenylarsonium pertechnetate at 11.09 μm.
No interferences from MoO_4^{2-}, WO_4^{2-}, Br^-, NO_3^-, SCN^-, SO_4^{2-}, PO_4^{3-}, IO_3^-,
UO_2^{2+}, and organic acids were observed, and MnO_4^- and ReO_4^- must be
absent. The sensitivity is in the microgram range.

4.5. Spectrophotometric Methods

Besides radiometric methods, spectrophotometric measurements in the
UV and visible region are the most employed methods for determining
Tc. TcO_4^- can be determined at 244 nm (ε = 6100 M^{-1} cm^{-1}, sensitivity = 1 μg/ml) [121]. In the visible region colored complexes
of Tc in its lower oxidation states with various inorganic and
organic ligands are used for determination. One of the most sensitive complexes is that formed by SCN^- and Tc(V) in acidic medium
(λ_{max} = 520, ε = 47500 M^{-1} cm^{-1}, sensitivity = 0.1 μg/ml) [127].
Other examples are the complex of Tc(IV) with 1,5-diphenylcarbo-
hydrazid (λ_{max} = 520, ε = 48600 M^{-1} cm^{-1}, sensitivity = 0.1 μg/ml)
[127] and $[Tc(OH)_3(CN)_4]^{3-}$ formed by dissolution of $TcO_2 \cdot 2H_2O$ in
NaOH/KCN (λ_{max} = 380, ε = 44000 M^{-1} cm^{-1}) [128].

4.6. Voltammetric Methods

Voltammetric methods have shown to be useful for quantitative determinations of Tc even in the presence of Mo and Re. In noncomplexing
neutral and alkaline media, the first wave ($E_{1/2}$ = -0.8 V versus SCE)
corresponds to a three-electron reduction [Tc(VII) to Tc(IV)]. The
current is diffusion-controlled and directly proportional to the Tc
concentration [129-131]. Depending on pH and/or the kind of com-
plexing agent in the supporting electrolyte, different reduction

steps (one to four electrons) for the first wave are observed, with half-wave potentials varying over a wide range. In most cases the first wave is followed by a second or third wave corresponding to reductions to lower oxidation states or by a catalytic wave. The sensitivity of direct voltammetric methods is in the microgram range. Anodic stripping voltammetry seems to be a sensitive method for determining TcO_4^- in alkaline solution. The potential of the hanging mercury drop electrode is held at -1.0 V (versus SCE) for a defined time and then swept toward positive values. Two peaks are observed (-0.33 and -0.20 V), both peak heights being directly proportional to the TcO_4^- concentration. This method allows the determination of 0.3 µg Tc in 20-ml solution [132].

Nuclear magnetic resonance and x-ray diffraction methods are used to elucidate structures of Tc compounds and the oxidation state of Tc in these compounds.

5. RADIODIAGNOSTIC USE OF TECHNETIUM

Nuclear medical in vivo diagnostic methods are based on the administration of a radioactive substance and the subsequent recording of its distribution in the body by measurement of the emitted radiation with a "scanner" or a computer-assisted "γ camera". If the radiopharmaceutical has been enriched in an organ, not only the size and position of the organ but also possible defects can be recognized by enhanced (hot spot) or reduced (cold spot) enrichment of activity. In the last years this static imaging of organs has been completed by dynamic measurements monitoring the uptake and excretion of the pharmaceutical as a function of time. This provides a view of the physiological function of the organ as well as its structure. A radionuclide suitable for nuclear in vivo diagnosis should fulfill the following features: (1) a short half-life so that even when a high starting activity is used, the radionuclide decays rapidly enough not to present a long-term hazard; (2) no emission of particles or very high energetic γ radiation to minimize the absorbed radiation dose to the patient;

(3) decay to a stable or very long-lived daughter nuclide to prevent long-term exposition; (4) emission of γ rays of sufficient energy to be easily detected through layers of tissue; and (5) ready availability in hospitals. The radiophysical properties of ^{99m}Tc largely correspond to these requirements.

The direct use of NaTcO$_4$ milked from a generator is limited to the imaging of the brain [133] and the thyroid [134]. The wide application of ^{99m}Tc radiopharmaceuticals relates to the fact that Tc in its lower oxidation states (III, IV, V) forms stable complexes with many suitable molecules. These radiopharmaceuticals can be categorized in two classes [135]:

1. The structure of the compounds of this class is not significantly altered by tagging with the nuclide and will show the same biological behavior as the untagged molecules. Examples of such Tc-tagged substances are:

   a. Particles and colloids: macroaggregated HSA and microspheres of HSA for evaluation of the lung arteriolar capillary bed [136]; minimicrospheres of HSA [137], tin hydroxide colloid, sulfur colloid, antimony sulfide colloid, and phytate for imaging the structure of liver and spleen [136].

   b. Proteins: HSA for blood pool assessment, lung and heart imaging [136,138], fibrinogen for thrombosis localization [139].

   c. Cells: erythrocytes for blood pool imaging, red blood cell mass determination, detection of vascular malformations, and nuclear cardiology [136,140].

   d. Small molecules: bone-imaging agents, e.g., polyphosphates [141], pyrophosphate [142], diphosphonate (MDP) [143], iminodiphosphonate.

2. In this class, the structure of the labeled compound differs from that of the parent molecule. The biodistribution of the compound depends on the properties of the complex. Examples are:

    a.   Kidney function agents: Tc-DTPA [144], Tc-EDTA [145], Tc-MIDA, Tc-citrate.

    b.   Kidney structure agents: Tc-gluconate, Tc-glucoheptonate [146], Tc-DMSA [147].

    c.   Infarct avid agents: Tc-pyrophosphate, Tc-HEDP [148], Tc-glucoheptonate [149], Tc-tetracycline [150].

    d.   Hepatobiliary agents: Tc-dihydrothiotic acid, Tc-HIDA [151], Tc-pyridoxylideneglutamate [152].

The radiopharmaceuticals given in this list are only examples of the most used ones. For clinical use, commercial kits of almost all common radiopharmaceuticals are available. These kits usually consist of a vial containing the substance to be labeled, sometimes a buffer and/or stabilizing agents, and a suitable reducing agent, mostly a salt of Sn(II), to obtain the lower oxidation state of Tc needed for complex formation. For stability reasons, the substances are often lyophilized and held in a nitrogen atmosphere. $NaTcO_4$ from a commercial generator is directly eluted with physiological saline into the vial. Without further chemical preparation, the radio-pharmaceutical is ready for use.

But even if the structure of the ligand present in large excess is well known, there are some difficulties in defining the final composition and structure of the administered ^{99m}Tc compound. It must be kept in mind that the quantity of ^{99m}Tc is extremely small; 1 mCi of ^{99m}Tc, a usual dosage, corresponds to 1.92 pmol (190 pg). Tin, usually present in much higher concentration, may compete with the reduced Tc in complex formation or is incorporated in the final complex as it could be proved for $Tc(DMG)_3(SnCl_3)OH \cdot 3H_2O$ [153]. However, this formation of mixed complexes seems not to be a general principle, e.g., ^{99m}TcHIDA does not contain Sn [135].

Another important question is the oxidation state of Tc in the administered complexes. Stable complexes are formed mainly by Tc(III), Tc(IV), and Tc(V). But in complexes even with the same ligand Tc may adopt different valency states depending on pH and amount of reductant present. Thus HIDA and DTPA contain Tc(III) at pH 4; at higher pH values the valency tends to be IV. As it could be shown for MDP, PP, and HEDP, different valency states may exist together [154]. Sometimes poor labeling of radiopharmaceuticals or an unexpected valency state of Tc results from lack of reducing agent. Although the initial ratio Sn/Tc being 10^2-10^5, by storage or faulty manipulations during labeling Sn(II) can be oxidized to Sn(IV). This can also occur when eluates from a generator not milked for days are used (see Sec. 2).

In many cases the final sterical structure and physiochemical state of Tc radiopharmaceuticals are unknown. However, it must be remembered that the aim of administering Tc-labeled pharmaceuticals is to investigate the structure or the function of an organ. The physiological and biochemical processes governing biodistribution and retention are not always fully understood. Therefore, in the past most work on ^{99m}Tc-labeled compounds has been done from the point of view of medical practice.

ABBREVIATIONS

bipy	bipyridyl
cat	1,2-dihydroxybenzene
cyclam	1,4,8,11-tetraazacyclotetradecane
diars	o-phenylene-bis(dimethylarsine)
diphos	bis(diphenylphosphine)ethane
DMG	dimethylglyoxime
DMSA	dimercaptosuccinic acid
DTPA	diethylenetriamine pentaacetate
EC	electron capture
EDTA	ethylenediamine tetraacetate
Et	ethyl
HEDP	hydroxyethylidene diphosphonate
HIDA	dimethylacetanilidoiminodiacetate
HSA	human serum albumin
IT	internal transition
MDP	methylene diphosphonate

Me	methyl
MIDA	methyliminodiacetic acid
nBu	n-butyl
Ph	phenyl
PP	pyrophosphate
SCE	saturated calomel electrode

REFERENCES

Recent reviews on technetium are:

a. *Gmelin Handbook of Inorganic Chemistry,* 8th ed.; Tc, Supplement Vol. 1, Springer-Verlag, Berlin, Heidelberg, New York, 1982.

b. M. J. Clarke and P. H. Fackler, The Chemistry of Technetium: Toward Improved Diagnostic Agents, in *Structure and Bonding, 50,* 57 (1982).

1. D. J. Mendelejeff, *Liebig Ann., Suppl., 8,* 205 (1871).

2. W. Noddack, I. Tacke, and O. Berg, *Berichte Berlin. Akad.,* 400 (1925).

3. W. Noddack, I. Tacke, and O. Berg, *Nature, 116,* 54 (1925).

4. C. Perrier and E. Segrè, *Nature, 140,* 193 (1937).

5. C. Perrier and E. Segrè, *Nature, 159,* 24 (1947).

6. P. Merril, *Astrophys. J., 116,* 21 (1952).

7. P. Merril, *Publ. Astronom. Soc. Pacific, 68,* 70 (1956).

8. J. Mattauch, *Naturwissenschaften, 25,* 738 (1937).

9. C. M. Lederer and V. S. Shirley, *Table of Isotopes,* 7th ed., John Wiley & Sons, New York, 1978.

10. M. Brucer, *J. Isotop. Rad. Technol., 3,* 1 (1965).

11. W. D. Tucker, M. W. Greene, and A. P. Murrenhoff, *Atompraxis, 8,* 162 (1962).

12. L. G. Stang, Jr. and P. Richards, *Nucleonics, 22,* 46 (1964).

13. L. G. Stang, Jr., W. D. Tucker, H. O. Baubs, Jr., R. F. Doering, and T. H. Mills, *Nucleonics, 12,* 22 (1954).

14. W. E. Winsche, L. G. Stang, Jr., and W. D. Tucker, *Nucleonics, 8,* 14 (1951).

15. P. Richards, *5th Congr. Nucl. Atti Uffic./Congr. Intern. Energia Nucl.,* Rome, 1960, Vol. 2, p. 253.

16. P. V. Harper, G. Andros, and K. Lathrop, *ACRH, 18,* 76 (1962).

17. F. D. Evans, in *The Atomic Nucleus*, McGraw-Hill, New York, 1955, p. 478.

18. H. A. Das, *Chem. Weekblad, 69,* 7 (1973).

19. M. Molter, *Chemiker Zeitg., 103,* 41 (1979).

20. T. D. Smith, J. R. Steiners, and P. Richards, *J. Nucl. Med., 16,* 570 (1975).

21. W. C. Porter, H. J. Dworkin, and R. F. Gutkowski, *J. Nucl. Med., 17,* 704 (1976).

22. L. G. Colombetti and W. E. Barnes, *Nuklearmedizin* (Stuttgart), *16,* 271 (1977).

23. E. Lebowitz, and P. D. Richards, *Sem. Nucl. Med., 4,* 257 (1974)

24. E. Shitata, *Nippon Aisotopu Kaigi Hobunshu, 12,* 152 (1976).

25. F. Nelson and K. A. Kraus, *Prod. Use Short-lived Radioisotop. React. Proc. Sem.,* Vienna, 1962, Vol. 1 (1963), p. 7.

26. L. G. Stang, Jr., *Prod. Use Short-lived Radioisotop. React. Proc. Sem.,* Vienna, 1962, Vol. 1 (1963), p. 3.

27. W. D. Tucker, M. W. Greene, and A. P. Murrenhoff, *Atompraxis, 8,* 163 (1962).

28. C. J. Fallais, *Rev. IRE Tijdschr., 3,* 4 (1978).

29. D. E. Troutner, E. L. Ferguson, and G. D. O'Kelly, *Phys. Rev., 130,* 1466 (1963).

30. O. P. D. Noronha, A. B. Sewatkar, R. D. Ganatra, C. K. Sivarama-krishnan, A. V. Jadhav, M. V. Ramaniah, and H. J. Glenn, *J. Nucl. Biol. Med., 20,* 32 (1976).

31. C. L. Ottinger, *Radioisotope Prod. Technol. Develop. Meeting,* Oak Ridge, Tenn., 1970, p. 34.

32. K. Motojima and M. Tanase, *Int. J. Appl. Rad. Isotop., 28,* 485 (1977).

33. M. Tanase and K. Motojima, *Int. J. Appl. Rad. Isotop., 32,* 353 (1981).

34. H. Arino and H. H. Kramer, *Int. J. Appl. Rad. Isotop., 29,* 97 (1978).

35. P. Reichold and H. P. Anders, *Radiochim. Acta, 5,* 44 (1964).

36. H. Ebihara and K. Yoshihara, *Bunseki Kagaku, 10,* 48 (1962).

37. W. Herr, *Z. Naturforschung, 7b,* 201 (1952).

38. V. Machan, M. Kalincak, and J. Vilcek, *Isotopenpraxis, 17,* 364 (1981).

39. C. M. E. Matthews and J. R. Mallard, *J. Nucl. Med., 6,* 404 (1965).

40. N. A. Morcos, G. A. Bruno, and T. A. Haney, at E. R. Squibb & Sons, Inc., U.S. Patent 4041317 (1977).

41. J. Cifka and P. Vesely, *Radiochim. Acta, 16,* 30 (1971).

42. R. G. Michell and K. Chapman, *Br. J. Radiol., 48,* 232 (1975).

43. W. C. Eckelman, G. Meinken, and P. Richards, *J. Nucl. Med., 13,* 577 (1972).

44. J. B. Gerlit, *Proceed. 1st Int. Conf. Peaceful Uses Atomic Energy,* Geneva, 1955, Vol. 7, p. 145.

45. O. Yoshinaga and K. Toyoaki, *Nippon Kagaku Zasshi, 85,* 128 (1964).

46. N. K. Baishya, R. B. Heslop, and A. C. Ramsey, *Radiochem. Radioanal. Lett., 4,* 15 (1970).

47. S. Tribalat and J. Beydon, *Anal. Chim. Acta, 8,* 22 (1953).

48. D. V. S. Narasimhan and R. S. Mani, *J. Radioanal. Chem., 33,* 81 (1976).

49. E. P. Belkas and D. C. Perricos, *Radiochim. Acta, 11,* 56 (1968).

50. G. D. Robinson, *J. Nucl. Med., 12,* 459 (1971).

51. C. Perrier and E. Segrè, *J. Chem. Phys., 5,* 712 (1937).

52. R. E. Boyd, *Radiopharm. Labeled Compounds Proc. Symp.,* Copenhagen, 1973, Vol. 1, p. 3.

53. L. G. Colombetti, V. Husak, and V. Dworak, *Int. J. Appl. Rad. Isotop., 25,* 35 (1974).

54. V. Machan, M. Kalincak, and S. Vilcek, *Isotopenpraxis, 17,* 364 (1981).

55. K. Kristensen, *13th Int. Ann. Meet. Soc. Nucl. Med.,* Copenhagen, 1975.

56. P. Vesely and J. Cifka, *Prep. Control Radiopharm. Conf.,* Vienna, 1970, p. 71.

57. H. Seiler, *Helv. Chim. Acta, 52,* 319 (1969).

58. P. Kleinert, E. Klose, and E. Schumann, *Isotopenpraxis, 16,* 92 (1980).

59. M. J. Buckingham, G. E. Hawkes, and J. R. Thornback, *Inorg. Chim. Acta, 56,* L 81 (1981).

60. J. G. Floss and A. V. Grosse, *J. Inorg. Nucl. Chem., 16,* 44 (1960).

61. J. C. Hileman, D. K. Huggins, and H. D. Kaesz, *Inorg. Chem., 1,* 933 (1962).

62. J. C. Hileman, D. K. Huggins, and H. D. Kaesz, *J. Am. Chem. Soc., 83,* 2953 (1961).

63. G. D. Michels and H. J. Svec, *Inorg. Chem., 20,* 3445 (1981).

64. W. Hieber, F. Lux, and C. Herget, *Z. Naturforschung, 20b,* 1159 (1965).

65. M. B. Cingi, D. A. Clemente, L. Magon, and U. Mazzi, *Inorg. Chim. Acta, 13,* 47 (1975).

66. F. Baumgärtner, E. O. Fischer, and U. Zahn, *Naturwissenschaften, 48,* 478 (1961).

67. C. Palm, E. O. Fischer, and F. Baumgärtner, *Tetrahed. Lett., 1962,* 253.

68. K. Schwochau and W. Herr, *Z. Anorg. Allgem. Chem., 319,* 148 (1962).

69. J. E. Fergusson and R. S. Nyholm, *Nature, 183,* 1039 (1959).

70. C. Orvig, Ph.D. thesis, Massachusetts Institute of Technology, 1981.

71. D. K. Huggins and H. D. Kaesz, *J. Am. Chem. Soc., 83,* 4474 (1961).

72. J. E. Fergusson and J. H. Hickford, *J. Inorg. Nucl. Chem., 28,* 2293 (1966).

73. G. Bandoli, D. A. Clemente, and U. Mazzi, *J. Chem. Soc., Dalton Trans.,* 125 (1976).

74. R. Münze, *Radiochem. Radioanal. Lett., 30,* 117 (1977).

75. J. Steigman, G. Meinken, and P. Richards, *Int. J. Appl. Rad. Isotop., 26,* 601 (1975).

76. F. A. Cotton and L. D. Gage, *Nouv. J. Chim., 1,* 441 (1977).

77. W. Preetz and G. Peters, *Z. Naturforschung, 35b,* 797 (1980).

78. F. A. Cotton, P. E. Fanwick, and L. D. Gage, *J. Am. Chem. Soc., 102,* 1570 (1980).

79. F. A. Cotton and G. Wilkinson, *Advanced Inorganic Chemistry,* 4th ed., John Wiley and Sons, New York, 1980, p. 883.

80. K. V. Kotegov, O. N. Pavlov, and V. P. Sheredov, *Adv. Inorg. Chem. Radiochem., 11,* 1 (1968).

81. J. Dalziel, N. S. Gill, R. S. Nyholm, and R. D. Peacock, *J. Chem. Soc., 1958,* 4012.

82. R. C. Elder, J. E. Fergusson, G. J. Gainsford, J. H. Hickford, and B. R. Penfold, *J. Chem. Soc. A., 1967,* 1423.

83. D. Brown and R. Colton, *J. Chem. Soc., 1964,* 714.

84. J. E. Fergusson and J. H. Hickford, *Aust. J. Chem., 23,* 453 (1970).

85. B. Gorski and H. Koch, *J. Inorg. Nucl. Chem., 31,* 3565 (1969).

86. A. Owunwanne, J. Marinsky, and M. Blau, *J. Nucl. Med., 18,* 1099 (1977).

87. F. P. Castronovo and R. C. Callahan, *J. Nucl. Med., 13,* 823 (1972).

88. R. D. Harcourt, *Speculations Sci. Technol.*, *2*, 527 (1979).

89. J. E. Turp, *Coord. Chem. Rev.*, *45*, 281 (1982).

90. D. Hugill and R. D. Peacock, *J. Chem. Soc. A.*, *1966*, 1339.

91. R. Colton and R. D. Peacock, *Quart. Rev. Chem. Soc.*, *16*, 299 (1962).

92. J. Jezowska-Trzebiatowska and M. Baluka, *Bull. Acad. Polon. Sci. Ser. Chim.*, *13*, 1 (1965).

93. F. A. Cotton, A. Davison, V. W. Day, L. D. Gage, and H. S. Trop, *Inorg. Chem.*, *18*, 3024 (1979).

94. W. Preetz and G. Peters, *Z. Naturforschung, 35b*, 1355 (1980).

95. G. Peters and W. Preetz, *Z. Naturforschung, 36b*, 138 (1981).

96. J. E. Smith, E. F. Byrne, and F. A. Cotton, *J. Am. Chem. Soc.*, *100*, 5571 (1978).

97. B. V. DePamphilis, A. G. Jones, M. A. Davis, and A. Davison, *J. Am. Chem. Soc.*, *100*, 5570 (1978).

98. A. G. Jones, B. V. DePamphilis, and A. Davison, *Inorg. Chem.*, *20*, 1617 (1981).

99. S. A. Zuckman, G. M. Freeman, D. E. Troutner, W. A. Volkert, R. A. Holmes, D. G. Van Derveer, and E. K. Barefield, *Inorg. Chem.*, *20*, 2386 (1981).

100. R. Münze, *Isotopenpraxis, 14*, 81 (1978).

101. M. E. Kastner, M. J. Lindsay, and M. J. Clarke, *Inorg. Chem.*, *21*, 2037 (1982).

102. H. S. Trop, A. G. Jones, and A. Davison, *Inorg. Chem.*, *19*, 1993 (1980).

103. A. F. Kuzina, A. A. Oblova, and V. I. Spitsyn, *Zh. Neorg. Khim.*, *17*, 2630 (1972).

104. W. de Kieviet, *Third Int. Symp. Radiopharm. Chem.*, St. Louis, 1980, p. 136.

105. B. V. DePamphilis, Ph.D. thesis, Massachusetts Institute of Technology, 1981.

106. C. D. Russell and A. G. Speiser, *J. Nucl. Med.*, *21*, 1086 (1980).

107. L. Ossicini, F. Saracino, and M. Lederer, *J. Chromatogr.*, *16*, 524 (1964).

108. H. Selig, C. L. Chernick, and J. G. Malm, *J. Inorg. Nucl. Chem.*, *19*, 377 (1961).

109. A. J. Edwards, D. Hugill, and R. D. Peacock, *Nature*, *200*, 672 (1963).

110. A. Guest and C. J. L. Lock, *Can. J. Chem.*, *50*, 1807 (1972).

111. K. Schwochau, L. Astheimer, J. Hauck, and H. J. Schenk, *Angew. Chem.*, *86*, 350 (1974).

112. E. Deutsch, W. R. Heineman, R. Hurst, J. C. Sullivan, W. A. Mulac, and S. Gordon, *J. Chem. Soc. Chem. Comm.*, *1978*, 1038.

113. H. Selig and J. G. Malm, *J. Inorg. Nucl. Chem.*, *25*, 349 (1963).

114. J. W. R. Dutton and R. B. Ibbett, AED Conf. 73-085-017, p. 1 (1973).

115. C. R. Walker, B. W. Short, and H. S. Spring, *Radiochem. Anal. Progr. Probl., Proc. 23rd Conf. Anal. Chem. Energy Technol.*, Gatlinburg, Tenn., 1979 (1980), p. 101.

116. R. A. Pacer, *Int. J. Appl. Rad. Isotop.*, *31*, 731 (1980).

117. M. E. Phelps, *Sem. Nucl. Med.*, *7*, 337 (1977).

118. S. Foti, E. Delucchi, and V. Akamian, *Anal. Chim. Acta*, *60*, 261 (1972).

119. G. E. Boyd and Q. V. Larson, *J. Phys. Chem.*, *60*, 707 (1956).

120. T. J. Anderson and R. L. Walker, *Anal. Chem.*, *52*, 709 (1980).

121. K. Schwochau, *Angew. Chem.*, *76*, 13 (1964).

122. G. E. Boyd, *J. Chem. Educ.*, *36*, 3 (1959).

123. J. W. Cobble, *Treat. Anal. Chem.*, *II, 6*, 412 (1964).

124. W. A. Hareland, E. R. Ebersole, and T. P. Ramachandran, *Anal. Chem.*, *44*, 520 (1972).

125. J. H. Kaye and N. E. Ballou, *Anal. Chem.*, *50*, 2076 (1978).

126. R. J. Magee and M. Al-Kayssi, *Anal. Chim. Acta*, *27*, 469 (1962).

127. H. Kupsch and M. Jovtscher, *Isotopenpraxis*, *15*, 1 (1979).

128. K. Schwochau and W. Herr, *Z. Anorg. Allgem. Chem.*, *318*, 198 (1962).

129. G. B. S. Salaria, C. L. Rulfs, and P. J. Elving, *Anal. Chem.*, *35*, 979 (1963).

130. R. J. Magee, I. A. P. Scott, and C. L. Wilson, *Talanta*, *2*, 376 (1959).

131. S. I. Zhdanov, A. F. Kuzina, and V. I. Spitsyn, *Zh. Neorg. Khim.*, *15*, 1567 (1970).

132. L. Astheimer and K. Schwochau, *J. Electroanal. Chem.*, *14*, 240 (1967).

133. D. A. Turner, A. A. Ali, M. G. Ochart, A. N. Sukerkar, E. W. Fordham, and G. V. S. Rayudu, *J. Nucl. Med.*, *18*, 258 (1977).

134. H. Nishiyama, in *Radiopharmaceuticals*, Vol. 2 (J. A. Sorenson, ed.), N.Y. Soc. Nuclear Medicine, 1979, p. 655.

135. H. D. Burns, P. Worley, H. N. Wagner, L. Marzilli, and V. Risch, *Chem. Radiopharm. Text Symp.*, Philadelphia, 1976 (1978), pp. 269/289.

136. S. C. Srivastava and P. Richards, in *Radiotracers for Medical Applications*, CRC Press, Boca Raton, Fla., 1981.

137. M. Villa, O. Pretti, R. Mosca, G. Plassio, and R. Pasqualini, *J. Nucl. Biol. Med., 20,* 168 (1976).

138. P. P. Benjamin, *Int. J. Appl. Rad. Isotop., 20,* 187 (1969).

139. M. Kirstein, R. Fridrich, and H. Seiler, *Thrombosis Res., 28,* 351 (1982).

140. W. Eckelman, P. Richards, W. Hauser, and H. Atkins, *J. Nucl. Med., 12,* 22 (1971).

141. G. Subramanian and J. G. McAfee, *Radiology, 99,* 192 (1971).

142. M. Kaye, S. Silverton, and L. Rosenthall, *J. Nucl. Med., 16,* 40 (1975).

143. G. Subramanian, J. G. McAfee, R. J. Blair, F. A. Kallfelz, and F. D. Thomas, *J. Nucl. Med., 16,* 744 (1975).

144. W. Eckelman and P. Richards, *J. Nucl. Med., 11,* 761 (1970).

145. S. Kato, K. Kurata, I. Ikeda, T. Sakoh, and H. Abe, *J. Nucl. Med., 14,* 415 (1973).

146. R. W. Arnold, G. Subramanian, J. G. McAfee, R. J. Blair, and F. D. Thomas, *J. Nucl. Med., 16,* 357 (1975).

147. D. Enlander, P. M. Weber, and L. V. dosRemedios, *J. Nucl. Med., 15,* 743 (1974).

148. N. D. Poe, *Sem. Nucl. Med., 7,* 7 (1977).

149. L. G. Lutzger and A. Alvi, *Sem. Nucl. Med., 6,* 83 (1976).

150. M. K. Dewanjee, C. Fliegel, S. Treves, and M. A. Davis, *J. Nucl. Med., 15,* 176 (1974).

151. M. D. Loberg and A. T. Fields, *Int. J. Appl. Rad. Isotop., 29,* 167 (1978).

152. E. Chiotellis, G. Subramanian, and J. G. McAfee, *Int. J. Nucl. Med. Biol., 4,* 21 (1977).

153. E. Deutsch, R. C. Elder, B. A. Lange, M. J. Vaal, and D. G. Lay, *Proc. Natl. Acad. Sci. USA, 73,* 4287 (1976).

154. C. D. Russell and A. G. Cash, *J. Nucl. Med., 20,* 532 (1979).

AUTHOR INDEX

Numbers in parentheses are reference numbers and indicate that an author's work is referred to although his name may not be cited in the text. Underlined numbers give the page on which the complete reference is listed.

SUBJECT INDEX

A

Absorption bands and spectra (and spectrophotometry, *see also* UV absorption spectra), 108, 157, 158, 163, 170, 171, 176, 187, 193, 195, 208-210, 214, 216, 230, 232, 340, 342

Acetate (or acetic acid)
  as ligand, 264
  buffer, 252
  trichloro-, 158, 163, 173, 240

Acidity constants, 254, 256, 262, 273, 274

Acrodermatitis enteropathica, 7, 8, 17

Actinides (*see also* individual names), 50, 61

Actinomycetes, 68

Adenosine monophosphate, *see* AMP

Adenosine 5'-triphosphate, *see* ATP

Adenylate cyclase, 30

Affinity constants, *see* Stability constants

Albumins, 96, 174, 175, 220, 292
  human serum, 292, 344
  pre-, 15

Alcohols (*see also* individual names), 158, 332
  aliphatic, 250
  allyl, 57
  amyl, 249, 250, 252, 257

Aldosterone, 140

Alizarin, 292

Alkali ions (*see also* individual names), 220

Alkaline earth ions (*see also* individual names), 86, 158, 220, 307

Alkaline phosphatase, 12

Alkaloids, 253

Allyl alcohol, 57

Alumina chromatography, 328, 330, 334

Aluminium(III), 31, 35-37, 39, 89, 107, 249, 254, 271, 284, 285
  clinical aspects, 213, 214
  determination, 213-221
  gel, 213
  standards, 118, 119

Alzheimer's disease, 214

Amines (*see also* individual names), 269
  butyl-, 269
  cyclic, 332
  diethyl-, 157

Amino acids (*see also* individual names), sulfur, 5, 11

p-Aminobenzoic acid, 262, 273

δ-Aminolevulinic acid, 186

Aminocarboxylic acids (*see also* individual names), 59-64, 95

Ammonia, 157
  as ligand, 265

AMP, cyclic-3',5'-, 238

Amyl alcohol, 249, 250, 252, 257

Anemia, 67
  Cooley's, 67
  sickle cell, 17

Anhydrase, carbonic, *see* Carbonic anhydrase

Animals (*see also* Mammal and individual species and names)
  abscess, 310
  tumors, 293

Antibiotics (*see also* individual names), 93, 262